Proceedings of the

3RD ASME INTEGRATED NANOSYSTEMS CONFERENCE
DESIGN, SYNTHESIS, AND APPLICATIONS

DEVICES AND SYSTEMS
NANOSCALE PHENOMENA
NANOMANUFACTURING
GENERAL TOPICS IN NANO SCIENCE AND ENGINEERING

presented at
3RD ASME INTEGRATED NANOSYSTEMS CONFERENCE
SEPTEMBER 22–24, 2004
PASADENA, CALIFORNIA USA

sponsored by
ASME NANOTECHNOLOGY INSTITUTE

A S M E
Three Park Avenue ◆ New York, N.Y. 10016

•• WELCOME FROM THE CHAIR ••

Dear Colleagues,

The ASME Nanotechnology Institute (NI) is four years old, and we have spent this time creating its infrastructure and its vision. I take this opportunity to thank the NI Advisory Board as well as ASME for their support.

The ASME NI has a dual mission: (i) to facilitate the development of nanotechnology and identify/establish the role of mechanical engineering, and (ii) to use this opportunity to modernize mechanical engineering so that it reflects the interdisciplinary nature of modern science and engineering, as well as the current/future needs and diversity of our society.

Over the last year, we have had some major developments, which I would like to share with you. We have created five committees within the ASME NI that cover the broad scope of our activities: Education; Nanomanufacturing; Devices and Systems; Nanoscale Phenomena; and Government, Venture and Social Impact. These committees have an open membership policy: I urge you to become members and participate in shaping the future. Note our website: http://nano.asme.org

The chairs and co-chairs have done an outstanding job in creating the ASME NI events, such as the NanoTraining Bootcamp, Integrated Nanosystems, the Nanotechnology Growth Opportunities for the Biotech and Medical Device Sectors Impact Forum, and the Nanomechanics: Sensors and Actuators Impact Forum. I take this opportunity to thank all chairs and co-chairs, as well as the participants, for making the last year an eventful and a very successful one.

This is the third annual Integrated Nanosystems Conference, and the interest and participation by the community has been spectacular. In contrast to the last two conferences, we have opened this one for many more unsolicited papers. The response has been so overwhelming that we had to create a large poster session to accommodate many of the papers. It is wonderful to see this level of interest.

In nanotechnology, the major challenge continues to be the design, synthesis, and integration of nanostructures to develop functional nanosystems. Our program has been designed to address this critical issue and provide you with potential breakthroughs and solutions.

While positive impact is certainly desirable, there could be many unintended consequences of nanotechnology that we should not ignore. This is an important issue and has become the topic of many discussions and forums. To directly address this issue, we have created a new panel that will discuss the health and environmental risks of nanotechnology.

The private investment community is extremely important in creating small businesses in nanotechnology, since they would probably be the engine for technological innovation, employment, and economic impact. This will be addressed through a panel session as well as a plenary lecture. For the time being, however, the federal government is perhaps the biggest and broadest investor in nanotechnology. The role of government in these early stages is vital, and this will be addressed through a panel session and a plenary lecture as well.

It is critical that the communication channels interconnecting industry, government, and academia remain open and clear. That is the goal of this conference. ASME has invested in the Nanotechnology Institute to provide you with solutions, and we hope that you will show your support by participating in our programs.

I truly think that we are experiencing a unique moment in history, where we can shape the development of a technology that will be ubiquitous and will positively impact many aspects our lives. We owe to our society to provide a clear vision of how and why this technology can and should be developed, and what impact it may have on all of us. The ASME Nanotechnology Institute will enable this to happen, and I hope you will join us to make it happen.

Sincerely,
Arun Majumdar
Chair, Advisory Board
ASME Nanotechnology Institute

Devices and Systems Committee
Chair: Chang-Jin (CJ) Kim
Co-Chair: Mark Shannon
Co-Chair: Ken Goodson

Education Committee
Chair: Pamela Norris
Co-Chair: Gang Bao
Co-Chair: Vinayak Dravid

Government/Venture/Social Impact Committee
Chair: Tom Kalil
Co-Chair: Tom Mackin
Co-Chair: Rosalyn Berne
Co-Chair: Kitu Bindra

Nanomanufacturing Committee
Chair: Margaret Blohm
Co-Chair: Steven Girshick
Co-Chair: S.V. Sreenivasan
Co-Chair: Vanita Mani

Nanoscale Phenomena Committee
Chair: Gang Chen
Co-Chair: Zhigang Suo
Co-Chair: Kenny Breuer

CONTENTS

NANO2004-46012

FABRICATION OF HYBRID BIONANODEVICES BASED ON COUPLED PROTEIN FUNCTIONALITY

D. HO, B.CHU, H. LEE, K. KUO, C.D. MONTEMAGNO

Bioengineering Department
University of California at Los Angeles, Los Angeles, CA 90095

ABSTRACT

Block copolymer-based membrane technology represents a versatile class of nanoscale materials in which biomolecules, such as membrane proteins, can be reconstituted. Among its many advantages over conventional lipid-based membrane systems, block copolymers can mimic natural cell biomembrane environments in a single chain, enabling large-area membrane fabrication using methods like Langmuir-Blodgett deposition, or spontaneous protein-functionalized nanovesicle formation. Based on this unique membrane property, a wide variety of membrane proteins possessing unique functionalities including pH/voltage gatable porosity, photon-activated proton pumping, and gradient-dependent production of electricity have been successfully inserted into these biomimetic systems.

INTRODUCTION

Energy transducing proteins remain at the forefront of the discussions of the applicability of biomolecules as a core technology for nanobiological devices in the areas of medicine, energy harvesting, and beyond. While conventional structural and mechanistic protein studies have yielded a plethora of information with respects to the fundamentals of protein behavior [1-4], our work is focused on the integration of these components towards the development of a fully biological fuel cell. As such, the ability to harness the efficiency and utility of these proteins in robust devices has necessitated the development of materials such as the biomimetic polymer membranes that can facilitate proper protein refolding while preserving protein function in the ambient environment. Our work utilizes the ABA triblock copolymer (PMOXA-PDMS-PMOXA) as the core membrane technology, which has previously been shown to support protein functionality [5-7].

The membrane protein, Bacteriorhodopsin (BR), found in Halobacterium Halobium, is a light-actuated proton pump that develops gradients towards the demonstration of coupled functionality with other membrane proteins to produce electricity through Bacteriorhodopsin activity-dependent reversal of Cytochrome C Oxidase (COX), found in *Rhodobacter Sphaeroides*. Using streptavidin–activated quantum dots interfaced with engineered protein constructs, we have previously demonstrated large-scale insertion of proteins into block copolymer Langmuir-Blodgett (LB) films as well as measurable pH changes based upon light-actuated proton pumping that are necessary for coupled protein functionality. Light actuated-activity across the protein-functionalized membrane when fully enclosed in a sol-gel matrix has also been observed using impedance spectroscopy. Demonstrations of resistance changes that were greater than $10k\Omega$ have provided indications of significant protein activity over time. More significantly, using spontaneously formed nanoscale polymeric vesicles (Fig. 1), measurements toward protein activity coupling between BR and COX using cyclic voltammetry and amperometric I-t analysis has demonstrated the potential of BR-driven electricity production by COX in the microampere range. Other biofuel cell testing and fabrication strategies will also be discussed.

MATERIALS AND METHODS

Protein Purification

Purple membrane purchased from Sigma was solubilized to 2mg/mL in 25mM phosphate buffer and 10% Triton X-100. Homogenized solution was then sonicated for 30 seconds and incubated at room temperature for 24 hours in dark room. The solution was then centrifuged at 110000 rpm for

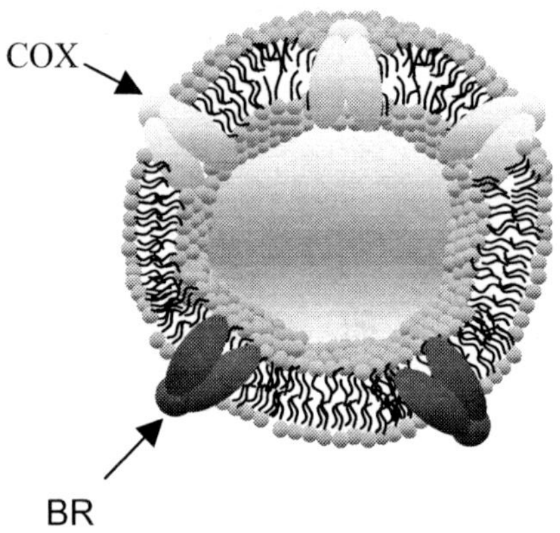

Figure 1. Depiction of hybrid protein-functionalized polymeric nanovesicle.

19 minutes, and only the supernatant was retained. The supernatant was then injected into HiPrep Sephacryl-100 column, preequilibriated with 8.12 mM Triton X-100 in liposome buffer with flow rate of 0.5 mL/min. All bacteriorhodopsin collected from the column was centrifuged at 110000 rpm overnight. The supernatent was discarded and the remaining bacteriorhodopsin pellet was centrifuged again.

Rhodobacter Sphaeroides was used as the over expression system in the form of a genetically engineered plasmid (15.9kb) for cytochrome c oxidase as well as its own circular chromosomes (>911kb, <2973kb). YZ100 produced COX that carried a 6-residue long histidine tag at the C terminus of the subunit 1, while 37-2 produced enzymes which have histidine tags at the subunit 2. The histidine tags were the key element used for Nickel NTA metal affinity purification since histidine binds with nickel. For the recombinant *Rhodobacter Sphaeroides*, the pH of Sistrom's media was sterilized and careful antiseptic methods of inoculation and handling were conserved to minimize contamination from other air-contained microbes [8].

Functionalized nanovesicle formation

Nanovesicles have previously been shown to successfully support protein refolding and functionality [9-11]. Based upon a simplified vesicle formation protocol, the block copolymer was solubilized to a 10-17 wt % concentration in ethanol and vigorously shaken overnight to result in a clear solution. Proteins were then added to the thoroughly-stirred polymer solution over several hours in order to induce integration between the proteins and polymer blocks. Additions were performed using amber vials in order to prevent BR activation. For measurements of BR-induced pH change across the membrane, the protein/polymer solution was then added to a vigorously-stirred polymer buffer solution (10mM Tris-HCl, pH 7.4) that contained a 100mM concentration of pyranine

(Molecular Probes). Nanovesicle samples for coupled protein functionality were added directly to the polymer buffer that did not contain pyranine. For polymer vesicles that contained both BR and COX, the proteins were added, as previously mentioned, in a slow fashion to induce optimum interfacing with the polymer blocks. TEM imaging (200kV) was used to confirm vesicle formation.

Lipid vesicles for coupled protein cyclic voltammetry were prepared using a conventional mixture of phosphatidylcholine, phosphatidic acid, as well as cholesterol to enhance membrane impermeability to ions [12]. The mixture was solubilized in diethyl ether as well as a lipid dialysis buffer (50mM Na_2SO_4, 50mM K_2SO_4, 2.5mM $MgSO_4$, 0.25mM DTT, 0.2mM EDTA, 20mM MOPS, pH 7.3). The solution was then inserted into a rotary evaporator (Buchi) and the pressure was reduced from atmospheric to 290 mbarr, at which point another 10ml of dialysis buffer was added to the evaporator flask. The resultant lipid solution was dialyzed against the dialysis buffer overnight in order to remove residual ether. Following dialysis, 1mg/ml solutions of BR and 3mg/ml of COX were introduced into the lipid solution and vortexed for proper mixing. This solution was further dialyzed against dialysis buffer to remove any excess detergent that could hinder protein insertion into the vesicles.

Cyclic voltammetry of functionalized vesicles

To perform cyclic voltammetry experiments, a commercially available analysis platform using a conventional three-electrode (reference, counter, working, Au) system (Genefluidics) was interfaced with an electrochemical work station equipped with cyclic voltammetry and amperometric detection capabilities (CH instruments) [13]. Controls containing no protein were added to the electrode (30μl) and scans were performed in the negative direction with a 0.2V sweep in 0.01V increments. Functionalized vesicle scans were performed both in ambient dark (BR-inactivated) and light (BR-activated) environments using the same scan parameters. Light incubation was performed using a 250W green-filtered light source for 1 minute durations. All samples contained a 1mg/ml tetramethylbenzidine (TMB) addition as a mediator.

RESULTS AND DISCUSSION

Protein Purification

BR purification was confirmed in the form of a 29kD band found through SDS-PAGE gel electrophoresis. The primary condition that was considered important for purification was to perform the chromatography in the dark so as to prevent premature activation of the protein.

COX elution was visually detectable as a distinctive green band in a metal affinity column. For the activity assays, NTA purification alone was considered good enough to produce functionally active COX with a higher yield.

Copyright © 2004 by ASME

<u>Functionalized nanovesicle formation</u>

Vesicle formation from polymers was confirmed using TEM spectroscopy. Figure 2a shows the formation of vesicles using the 4METH derivative while figure 2b shows the formation of the 8METH vesicle. It was noticed that significantly more 8METH vesicles (>50, 5µl) were formed over the shortchain version (~10, 5µl) which was attributed to the decreased stability seen in the 4METH polymer which was previously confirmed through LB compression trials [11]. Following filtration of the vesicles with 200µm diameter pore sizes, the vesicles were shown to possess similar diameters as well. Lipid vesicle formation was confirmed through protein functionality assays of BR and COX [12]. In this case, BR was inserted into the lipid solution and tests were performed to determine light-activated BR activity as protons are pumped across the membrane to interact with the pH-sensitive pyranine dye in the core of the vesicle. Light-dependent pH changes were seen in both the polymer and lipid vesicles (Fig. 3,4), which were further used to ascertain that the vesicles had indeed formed and possessed the proper structure. Measurement of pH change activity across the polymer vesicle controls revealed little to no pH change (±0.03) in both the light and dark conditions. Protein-functionalized vesicles, when incubated with green-filtered light, were able to consistently produce fluorometric peaks across the 402 and 456 wavelengths that corresponded to a pH change of at least ~0.18-0.2 (Fig. 3a). However, it was seen that when the vesicle endgroups were crosslinked, the resulting pH change rose to ~0.5 (Fig. 3b). Furthermore, it was seen that the pH changes occurred very rapidly, which was attributed to the increased permeability seen with copolymer vesicles due to polymer molecule rigidity, etc. Lipid vesicle fluorometric changes as well as pH changes were observed across time intervals of approximately ten minutes.

<u>Cyclic voltammetry of functionalized vesicles</u>

Following the incubation of the polymer control sample that contained only nanovesicles with the electrodes in both light and dark samples, scans revealed a baseline value for current measurements (Fig. 5). It could be seen that a clearly visible redox curve or peak was absent in the control scan as well, indicating that without protein activity, there was no current production in the background reading.

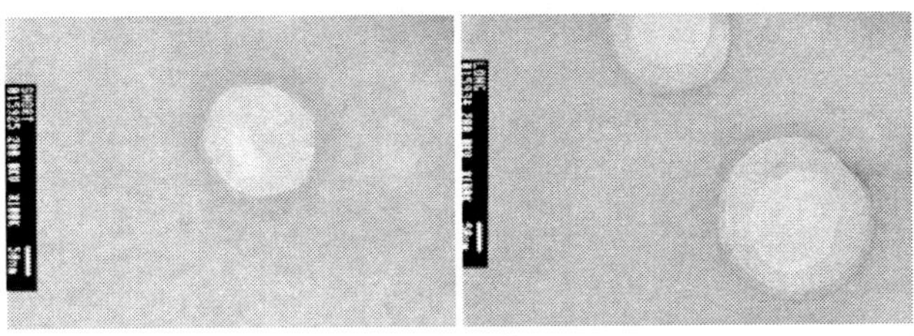

Figure 2. Figure 2a shows the formation of a 4METH nanovesicle while figure 2b shows the formation of 8METH vesicles.

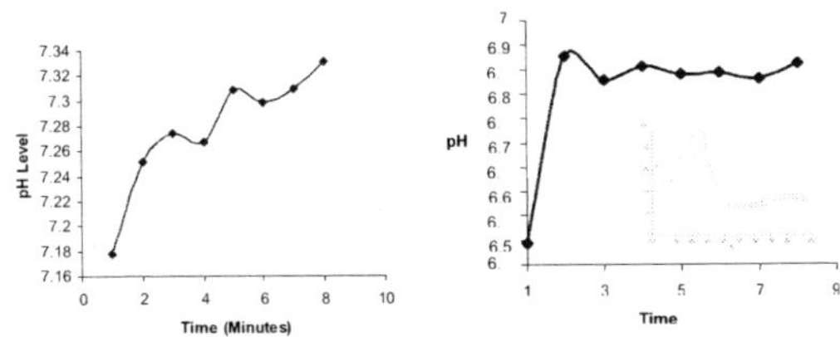

Figure 3. Figure 3a represents pH change from BR-driven proton pumping in uncrosslinked vesicles. Figure 3b represents a pH change of ~0.5 due to light-driven BR activity in crosslinked vesicles. (Inset: Fluorometric change corresponding to pH change of ~0.5).

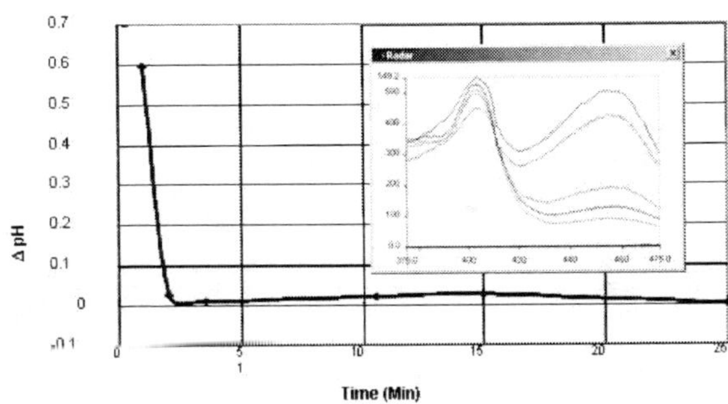

Figure 4. Light-dependent pH change in lipid vesicles from BR-driven activity. (Inset, Fluorometric change corresponding to pH change of 0.6)

After the protein-functionalized nanovesicles were interfaced with the electrode, current production values increased for both the light and dark-incubated samples as seen in the presence of clear redox peaks (Fig. 6). It is believed that a redox peak was present for the dark-incubated hybrid sample due to the fact that there was possible exposure to some ambient light that could have accounted for the increased current density when expected performance indicates that inactivated BR in the dark should produce no current. However, when comparing light-activated BR/COX vesicles with dark-incubated BR/COX vesicles as well as the initial controls, it could clearly be seen that light-activated samples possessed higher current densities that equated to approximately 12.5µA/cm^2 while controls resided in the nA range.

Amperometric current vs. time plots were also obtained for the samples. Fig. 7 demonstrates the production of current using hybrid functionalized polymeric vesicles measured against the control, which produced no current. Amperometric measurement was also conducted for samples that contained only one type of protein, either BR or COX, to determine whether or not each protein alone could be attributed to current production. These measurements resulted in no current being produced, demonstrating a preliminary indication that single protein activity does not result in significant current production from vesicles. Careful attention was paid when setting the scan parameters to set upper and lower boundary limits that were not near the hydrolysis point of water.

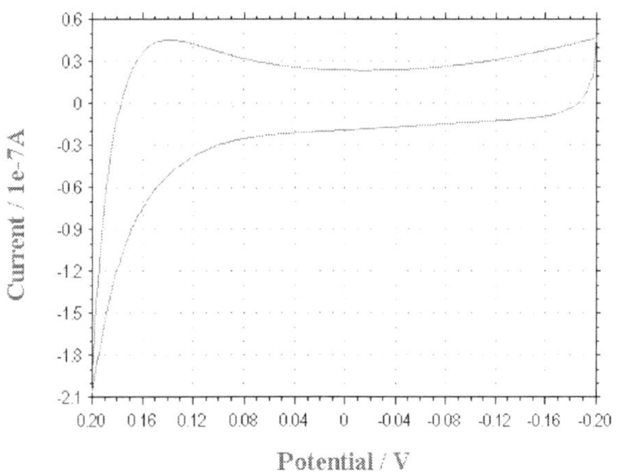

Figure 5. Background value for control polymer vesicles (no protein).

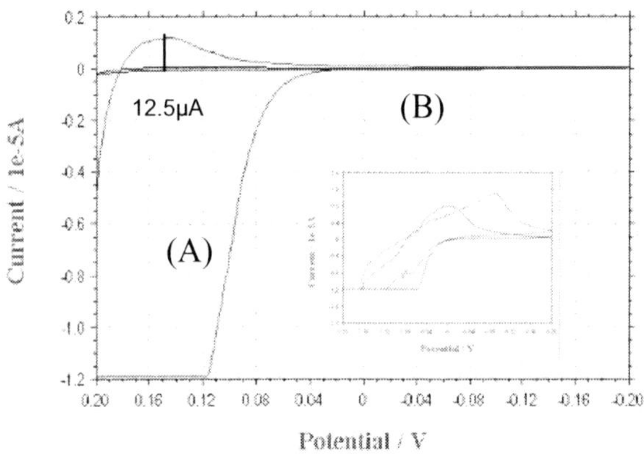

Figure 6. Measurement of hybrid vesicle current (A) against control current (B) reveals microampere-range currents for functionalized vesicles. (Inset: Redox peaks are visible for both light and dark samples, though incubation in light results in increased current readings.)

Conductivities of the electrodes used were calibrated and shown to be nearly identical in value that ruled out varying electrode conductivities as a reason for the increased current densities seen in hybrid vesicles. Vesicles are believed to be an efficient approach towards the development of electricity from fully biological unit cells as the nanoliter environments enclosed by the nanovesicles represent the ideal controlled environment for maintaining the light-driven BR proton gradients that are necessary to effect the coupled functionality with the COX for electron release from water. The vesicles are very stable, as they can undergo centrifugation while still maintaining structural integrity. Furthermore, vesicles were very robust as they could sustain protein functionality and spherical conformations for a minimum of several weeks.

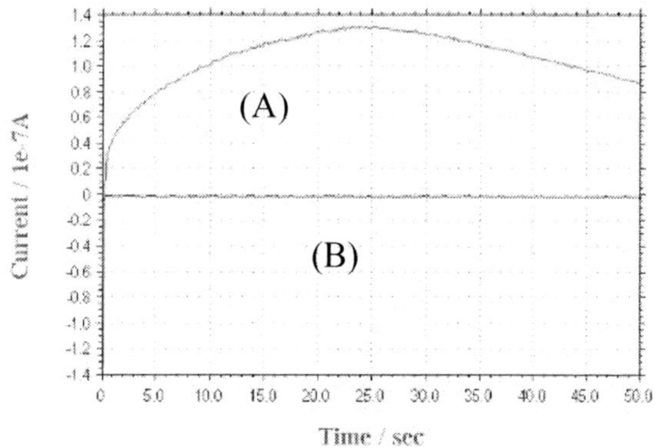

Figure 7. Amperometric detection of current shows an increase in current production (A) over protein-only controls (B).

Cyclic voltammetry was also performed using lipid vesicles. Controls that contained no protein were interfaced with the same electrodes used for the polymer vesicle measurements, and the scan parameters were preserved for lipid assays as well. The resultant CV plot indicated that there was no current produced for this reaction given by the absence of a clearly visible redox peak (Fig. 8). Vesicles containing both proteins were then added to the electrode and quickly scanned in a dark environment, indicating the presence of a small redox peak corresponding to ~ 1.5µA. The sample was then incubated with a green-filtered light source for 1 minute, and the subsequent scan revealed a peak corresponding to values just above 3µA. Similar to the polymeric vesicles, it was believed that redox peaks were still observed in dark-incubated samples due to the fact that there was some light exposure prior to green-light incubation which resulted in increased current densities. Furthermore, it should be mentioned that BR is not only sensitive to green light, but ambient light as well that could prematurely activate the protein to cause current increases. Fig. 9 represents an overlay of dark-incubated hybrid vesicles, light-incubated vesicles, as well as protein-only controls. From this it can be observed that that hybrid samples possess significantly higher current outputs over the backgrounds obtained from the control samples.

Measurement of membrane protein functionality can be achieved by using several conventional methods of protein refolding/activity assays [14]. Our work seeks to integrate individual cellular components towards the development of biomolecules-based devices. As such, conventional membrane protein activity detection methods, which typically take place across $1mm^2$ areas are not deemed to be sufficient for device fabrication methods.

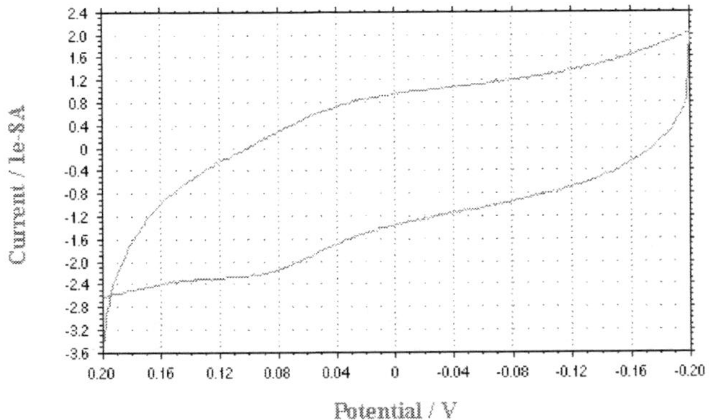

Figure 8 Lipid vesicle control reveals the lack of redox peaks.

Towards the production of a hybrid bionanodevice that is capable of producing useable amounts of electricity, the fabrication strategy is centered at the core technology of demonstrating a coupled reaction between the BR and COX, whereby the green light-driven activity of BR results in the development of a proton gradient used to drive COX in reverse, releasing electrons. This technology is extended to the fact that the gradients developed by the proteins must be sustained by the membrane systems in order to preserve

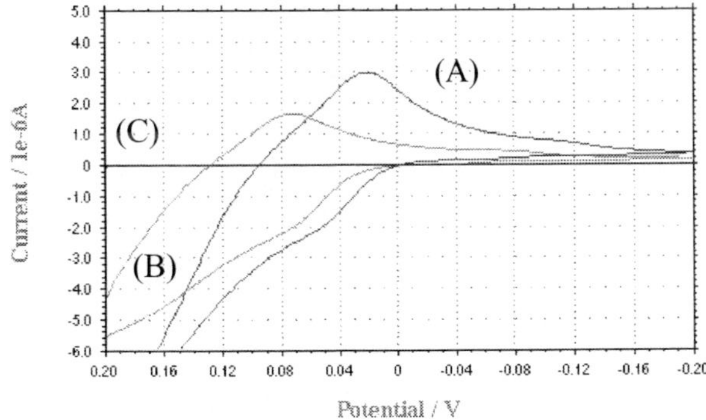

Figure 9. A comparison of all three plots shows an increased current for light-incubated hybrid samples (A) when compared with dark-incubated samples (B), as well as control samples (C).

device performance. As such, the fabrication of the devices must take into account the optimal method of protein reconstitution along with membrane integrity. Planar fabrication methods including Langmuir-Blodgett deposition are very useful for large-area development of devices that contain pore proteins for filtration, biosensing, etc due to the fact that the fluid volumes on both sides of the membrane, which can reach very large values depending on the area used (millileter-liter range) are irrelevant to device performance. However, in cases where gradients are being developed by proteins, it is unrealistic to assume that global properties such as pH can be solely manipulated by proteins when fluid volumes in the milliliter to liter range exist on either side of the membrane. Previously reported data demonstrating BR-based pH change occurred in the nanoliter environments of vesicles, which is substantially more easily achieved than planar conformations. As such, development of a biofuel cell based upon integrating the proteins with vesicles is a more feasible approach. Device output can be regulated based upon vesicle concentration. Furthermore, three-dimensional patterning of the liquid vesicle-containing solutions can be achieved through the casting of the vesicles in a tetramethylorthosilicate (TMOS) sol-gel matrix (Fig 10). Previous studies of protein incorporation into sol-gels have indicated that functionality can be enhanced due to the pore stability seen with the TMOS network. Therefore, packaging the final device based upon the harvesting of electrons from this nanovesicular system integrated with sol-gel encapsulation delineates a compelling approach towards using a fully-biological interactive system to derive electricity from light energy.

For both the polymer and lipid vesicles, studies have been conducted to assess the functionality of BR as well as COX retaining functionality individually in the membrane systems. Based upon the measurement of BR-dependent proton transport in the polymer and lipid systems, preservation of its structure and functionality was demonstrated [15]. COX-dependent oxygen-depletion as well as pH-dependent reduction of the heme a group in the lipid systems [16] as well as Cytochrome c- based COX activity in planar polymer impedance assays [17] were able to demonstrate the preservation of COX functionality. As such, the measurement of current increases produced in vesicles that possess both protein compared to control vesicles that do not have current increases delineates a promising step towards the demonstration of coupled protein functionality.

CONCLUSIONS

The reconstitution of energy conversion biomolecules into biocompatible block copolymer systems has realized the utility of membrane proteins towards the fabrication of fieldable integrated devices. Beyond the study of basic biology behind membrane proteins, our work has delineated the feasibility of coupling protein functionality for practical purposes. These accomplishments provide considerable

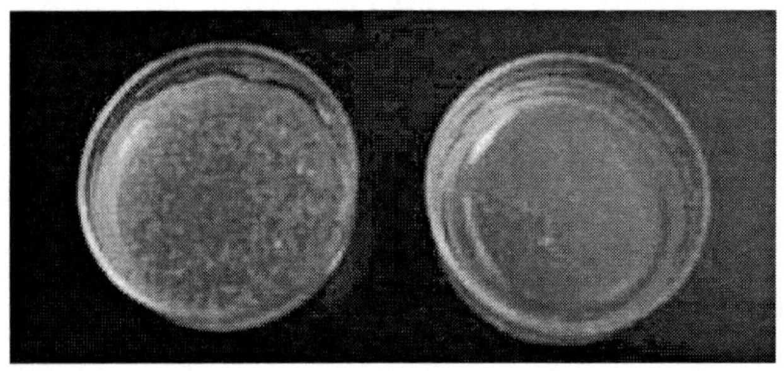

Figure 10. A tetramethoxyorthosilicate (TMOS) sol-gel matrix was used to cast the hybrid vesicle solution to enable three-dimensional patterninig. Left: 2X conc. SolarGel; Right: 1X conc. SolarGel

promise toward the production of a fully-biological integrated device that is based upon individual molecular function. Utilizing protein-functionalized polymer and lipid-based vesicles, cyclic voltammetry trials revealed the production of $\mu A/cm^2$-range current densities based upon systems that contained both proteins, visible from redox peaks. Control experiments that did not contain protein produced substantially lower background current density readings in the pA/cm^2 to nA/cm^2 range. Furthermore, this work has identified a key fabrication strategy that can be used to develop a robust biofuel cell that has possesses a vesicle concentration-dependent output as well as a stable packaging configuration. Future work will concern the generation of mA current densities based on BR-dependent COX reversal, along with sol-gel packaging of the hybrid vesicles for 3-D patterning.

ACKNOWLEDGMENTS

The authors would like to acknowledge Na Li for CV guidance, as well as Hyo-jick Choi for TEM assistance. This work was funded by DARPA.

REFERENCES

1. Spudich, J. L., Bogolmolni, R.A. "The mechanism of color discrimination by a bacterial sensory," *Nature* vol. 312, 509-513, (1984).
2. Spudich, J.L., Bogolmolni, R.A. "Sensory rhodopsin in halobacteria," *Annu. Rev. Biophys. Chem.*, vol.17, 193-215, (1984).
3. Steinberg, G., Friedman, N., Sheves, M., Ottolenghi, M. " Isomer composition, and spectra at the dark and light adapted forms at artificial bacteriorhodopsin," *Photochem. Photobiol.*, vol.54, 969-976, (1991).
4. Stoeckenius, W., "From membrane structure to bacteriorhodopsin," *J. Membr. Biol*, vol.. 139 139-148, (1994).
5. Meier, W., Nardin, C., Winterhalter, M., "Reconstitution of Channel Proteins in (Polymerized) ABA Triblock Copolymer Membranes," *Angew. Chem. Int. Ed.*, vol. 39, pp 4599 4602, (1999).
6. Winterhalter, M., Hilty, C., Bezrukov, S.M. , Nardin, C., Meier, W., Fournier, D., "Controlling Membrane Permeability with Bacterial Porins: Application to Encapsulated Enzymes." *Talanta* 55:965-971, (2001).
7. Nardin, C., Widmer, J., Winterhalter, M., Meier, W., "Amphiphilic Block Copolymer Nanocontainers as Bioreactors." *Eur. Phys. Journ. E.* 4:403-410, (2001).
8. Zhen, Y., Qian, J., Follmann, K., Hayward, T. , Nilsson, T., Dahn, M., Hilmi, Y., Hamer, A.G. , Posler, J.P., Ferguson-Miller, S., *Protein expression and purification* 13, 326-336, (1998).
9. Nardin, C., Winterhalter, M., Meier, W., "Giant Free-Standing ABA Triblock Copolymer Membranes," *Langmuir* vol.16, pp 7708-7712, (2000).
10. Winterhalter, M., Klotz, K., Benz, R., in *Electromanipulation of Cells*, edited byU. Zimmerman, G.A. Neil ,CRC Press, Boca Raton, p.137,(1995).
11. Ho, D., Chu, B., Schmidt, J., Brooks, E., Montemagno, C.D., "Hybrid Protein/Polymer Biomimetic Membranes" accepted to *IEEE Trans. Nanotechnology* June, (2004).
12. Lee, H., Ho, D., Schmidt, J., Montemagno, C.D., " Reconstitution of energy converting proteins in biocompatible materials," *IEEE Proc. Nanotechnology,* vol. 2, 733-736, (2003).
13. Gau, J., Lan, E., Dunn. B., Ho, C.M., Woo, J., " A MEMS based amperometric detector for E. coli bacteria using self-assembled monolayers," Biosens. Bioelec., vol. 16, pp. 745-755, (2001).
14. Hazard, A., Montemagno, C.D., "Improved purification for thermophilic F1F0 ATP synthase using n-dodecyl b-D-maltoside," *Arch. Biochem. Biophys.*, vol. 407, 117-124, (2002).
15. Ho, D., Chu, B., Lee, H., Montemagno,C.D., "Block Copolymer-Based Biomembranes Functionalized with Energy Transduction Proteins," *Proc. Mat. Res. Soc.,* vol.9, April 12-16, 2004
16. Ho, D., Chu, B., Lee,H., Montemagno,C.D., "Nanoscale Protein/Polymer Functionalized Materials," *SPIE Proc. Smart Materials/Nanotechnology*, vol. 11, March 15-18, 2004
17. Ho, D., Chu, B., Lee, H., Montemagno, C.D., "Protein-driven Energy Transduction Across Polymeric Biomembranes," submitted to *Nanotechnology*, March, 2004

NANO2004-46013

MULTIWALLED CARBON NANOTUBE / NANOFIBER ARRAYS AS CONDUCTIVE AND DRY ADHESIVE INTERFACE MATERIALS

Tao Tong[*], Yang Zhao[+], Lance Delzeit[†], Ali Kashani[+], and Arun Majumdar[*§]

[*] Department of Mechanical Engineering, UC Berkeley, Berkeley, CA 94720
Materials Sciences Division, Lawrence Berkeley National Laboratory, Berkeley, CA 94720
[+] Atlas Scientific, Inc., San Jose, CA   95120
[†] NASA Ames Research Center, Moffett Field, CA 94035

ABSTRACT

We demonstrate the possibility of making conductive and dry adhesive interfaces between multiwalled carbon nanotube (MWNT) and nanofiber (MWNF) arrays grown by chemical vapor deposition with transition-metal as catalyst on silicon substrates. The maximum observed adhesion force between MWNT and MWNF surfaces is 3.5 mN for an apparent contact area of 2 mm by 4 mm. The minimum contact resistance measured at the same time is ~20 Ω. Contact resistances of MWNT-MWNT and MWNT-gold interfaces were also measured as pressure forces around several milli-Newton were applied at the interface. The resulting minimum contact resistances are on the same order but with considerable variation from sample to sample. For MWNT-MWNT contacts, a minimum contact resistance of ~ 1 Ω is observed for a contact area of 2 mm by 1 mm. The relatively high contact resistances, considering the area density of the nanotubes, might be explained by the high cross-tube resistances at the contact interfaces and limited interpenetration of the nanotube arrays.

INTRODUCTION

Electrical contacts are of key importance to the quality and reliability of micro-electronics. For example, in radio frequency (rf) micro-electro-mechanical (MEMS) switches, low insertion loss requires low contact resistances and continuous operation requires repeatable and reliable contact characteristics [1]. Electrical contact resistances arise from the macroscopic lack of planarity, micro-roughness of the mating surfaces, as well as the contact characteristics itself. When two surfaces are brought into contact with each other, the actual contact area may be much smaller than the apparent contact area depending on the surface roughness, elastic and plastic properties of the surface, as well as the applied load at the interface. Typical metal-metal contacts, e.g., gold, have sub-ohm resistances [1,2]. But metal contacts are often limited by contact surface degradation and local melting of the contact area (in power switch applications) [1-3]. As a potential solution to the interface problem, carbon nanotubes represent a class of materials possessing extraordinary electrical, mechanical, and thermal properties [4]. While single-walled carbon nanotubes (SWNTs) can be metallic or semiconducting depending on chirality, multiwalled carbon nanotubes (MWNTs) and multiwalled carbon nanofibers (MWNFs) [5] are usually metallic. Conductive nanotubes have relatively low through-tube resistances and high current carrying capabilities. They are mechanically sturdy and do not readily fatigue. Thus acting as an interface bridging material, carbon nanotubes might effectively improve the characteristics and reliability of the electrical contacts. Carbon nanotubes are also excellent thermal conductor with through-tube thermal conductivity around 3000 W/m-K at room temperature [6], and thus also have the potential of being used as thermally conductive switches. Recently, chemical vapor deposition (CVD) with the aid of transition-metal as catalyst has emerged as a routine method for growing relatively large quantity of carbon nanotubes [7]. Compared with laser ablation and carbon arc processes, CVD method gives the flexibility of growing carbon nanotubes on patterned surfaces with good vertical alignment. This makes growing carbon nanotube-padded electrical contacts a possibility.

Electrical transport properties of individual carbon nanotubes have long been a focus of research since the invention of carbon nanotubes. White and Todorov [8] showed

[§] Corresponding author: Tel: 1-510-643-8199 Email: majumdar@me.berkeley.edu that for metallic (armchair) SWNTs, when the contacts are perfect and transport is ballistic (tube length < ~10 μm), the resistance through the nanotube is theoretically 6.5 kΩ, half of the quantum resistance $h/2e^2$ (two open conduction channels). Soh et al. [9] experimentally measured their SWNT bridging two electrode islands with Ti/Au end contacts had a minimum resistance of 20 kΩ. For multiwalled carbon nanotubes (MWNTs), de Pablo et al. [10] showed that with good contacts formed by Ti/Au covering and in the ballistic range, the MWNTs have quantized fraction of the quantum resistance (multiples of quantum conductance) scaling with the layers of the individual MWNTs (open tip). Frank et al. [11], however, reported the resistances of individual MWNTs on the order of quantum resistance without scaling with the number of graphite layers. They attribute this to the close end of the MWNT tips and large interlayer resistances for their almost defect-free arc produced nanotubes. Biswas et al. [12] reported using a conductive tip atomic force microscope (CT-AFM) to tap the individual MWNTs deposited into porous anodized alumina template with platinum back contact, the resistivity of individual nanotubes is 1.4×10^{-5} Ω·m and can sustain a current density larger than $\sim 10^5$ A/cm^2. Li et al. [13] showed a measured individual tube contact resistance of ~300 kΩ for their CVD deposited MWNF array (they were calling them MWNTs) on silicon substrate.

From the mechanical perspective, the vertically aligned MWNT arrays often have curly surface structures with tube tips entangling with each other and the MWNF arrays are often thicker in diameter, shorter, and straighter. Adhesion force between two such surfaces could thus be generated during contact due to the mechanical entanglement and van der Waals forces like a 'nano-velcro'. Berber et al. [14] showed through their total energy and molecular dynamic calculations, a large force of 3 nN is needed to disengage two entangled nanotube 'hooks'. To the authors' knowledge, no experimental demonstrations have been shown yet in literature.

The current study focuses on the macroscopic behavior of the electrically conductive and adhesive characteristics of the contact interfaces formed by MWNT and MWNF arrays. Comparisons are made with the theoretical predictions and previous experimental work on individual nanotubes. With the development of local synthesis of carbon nanotubes [15], our study demonstrates the potential of applying MWNTs/MWNFs as conductive and dry adhesive interface material that can be used in electrical switches or other device applications.

EXPERIMENTS

In this work, the MWNTs and MWNFs are grown by thermal CVD and plasma enhanced CVD with transition-metal catalyst, respectively [5,7]. Highly Boron doped (10^{19} /cm^3) silicon wafer is used as the substrate material. First, a layer of Al (10 nm) followed by a catalyst layer of Fe (10 nm) is first deposited onto the silicon substrate by ion beam sputtering. The metals used in the experiments are 99.9+% pure and are

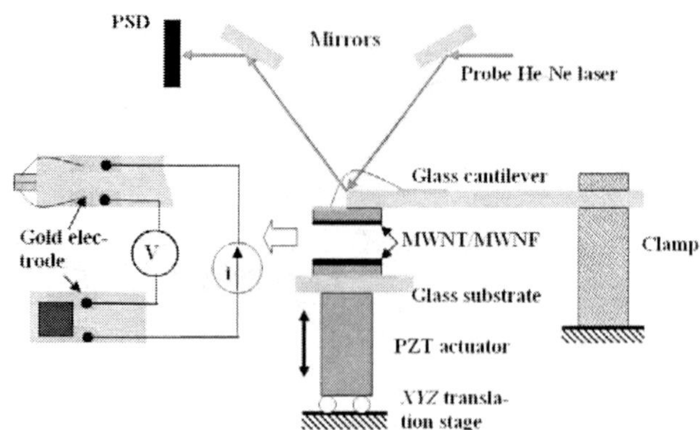

Fig. 1 Experiment schematics of the contact force and electrical resistance measurement.

sputtered using a VCR Group Incorporated Ion Beam Sputterer (model IBS/TM200S). In the homemade CVD reactor, argon (Scott Specialty Gases, 99.999% pure, 1000 sccm) is used to purge the reactor while the furnace is heated up to 750 °C. After the furnace temperature reaches 750 °C, it is allowed an additional 10 min for the temperature to equilibrate. The gas flow is then switched to 1000 sccm of ethylene (Scott Specialty Gases, 99.999% pure) for 10-min growth time. The gas flow is switched back to argon after the growth to purge ethylene and prevent back flow of air into the reactor. The furnace is allowed to cool below 300 °C before exposing the nanotubes to air. For the carbon nanofibers growth, the feedstock gases are usually a combination of 10-20% methane and 80-90% hydrogen at a total flow rate of 100 sccm. The substrate is heated up to 900 °C. An inductively coupled plasma reactor operated in a capacitively coupled mode was used to facilitate the nanofibers growth. Detailed discussions on nanotube and nanofiber growth are seen in Ref. [5,7].

The key factor that differentiates carbon nanofibers from nanotubes is that the multiwalled carbon nanotubes have a nested Russian doll structure with the graphitic basal planes parallel to the tube axis, while the nanofibers' graphitic basal planes are to a non-zero (up to 30°) angle to the tube axis forming a stacked-cone structure along each fiber [5]. Further, the nanotubes are base-growth mechanism with catalyst particles at the bases, and the nanofibers are tip-growth mechanism with catalyst particles at the tips. Under the growth conditions in this work, the multiwalled nanotubes have diameters ranging from 10-30 nm with the tube density estimated to be $\sim 10^{10}$ tubes/cm^2 and nanotube tower heights of ~100 μm for a 10 min growth time. The nanofibers usually have larger diameters ranging from 40 to 90 nm with the tube density estimated to be $\sim 10^8$-10^9 fibers/cm^2, and for the same 10 minute growth time the heights of the fibers are usually on the order of several micron with larger variation within the array than the nanotubes.

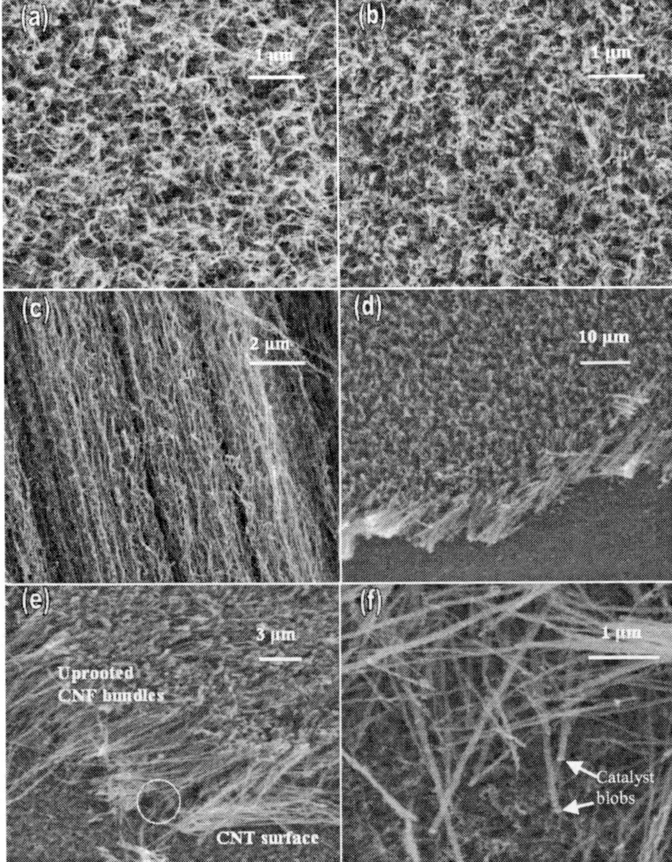

Fig. 2 SEM pictures of (a) fresh carbon nanotube array surface, (b) nanotube array surface after contact experiment, (c) side view of the nanotube bundles, (d) fresh nanofiber array surface, (e) part of the nanotube sample surface after contact experiment showing some nanofiber bundles being uprooted from their own substrate and sticking into the nanotubes' curly surface, (d) magnified view of the circular region in (e) showing nanofibers' penetration into the nanotubes' surface.

The interaction force and the contact resistance are measured in an experimental setup schematically shown in Fig. 1. The upper sample is loaded facing downward at the free end of a glass or quartz cantilever. The lower sample is loaded on a glass substrate facing upward. The glass substrate sits on a piezo actuator (PZT material), which is mounted on a three dimensional orthogonal manual translation stage. The lower stage can be brought up and down manually or by actuating the PZT. Upon contact, the deflection of the cantilever is monitored by a probe laser shooting at the tip of the cantilever and monitoring the displacement of the reflected laser spot on a position sensitive device (PSD). The manual translation stage has a readable resolution of 10 µm in the vertical direction. The piezo actuator in our experiment can deform vertically within a 6.8 µm range continuously with a dc driving voltage from 0 V to 100 V. Since the hysteresis of piezo materials depends on loading situations, in our experiment the voltage ramping speed is always kept constant at 2 V/sec from 100 V to 0 V to improve the precision and repeatability. The piezo actuator was calibrated through a Michelson interference experiment. The obtained displacement-voltage relation was compared with PSD response curve with good agreement when the lower stage and the cantilever are made into close contact and move together. Thus, sub-micron precision can be obtained for the piezo actuator.

The spring constant of the cantilever is determined through the resonant frequency of the cantilever. The fundamental frequency of the cantilever can be measured by observing the impulse response of the cantilever. Theoretically, the fundamental frequency f_1 of a rectangular plate cantilever neglecting the end mass can be expressed as [16]:

$$f_1 = \frac{1}{2\pi} \frac{3.6637}{L^2} \left(\frac{EI}{\rho ab} \right)^{1/2} \qquad (1)$$

where L is length of the cantilever, E the Young's modulus, I the moment of inertia, ρ the density of the cantilever material, a the width of the cantilever, and b is the thickness of the cantilever. For the free end cantilever, the transverse spring constant at the tip is:

$$k = \frac{3EI}{L^3} \qquad (2)$$

Rearranging Eqs. (1) and (2), one can express the spring constant in terms of the fundamental frequency, material density, and the beam geometry:

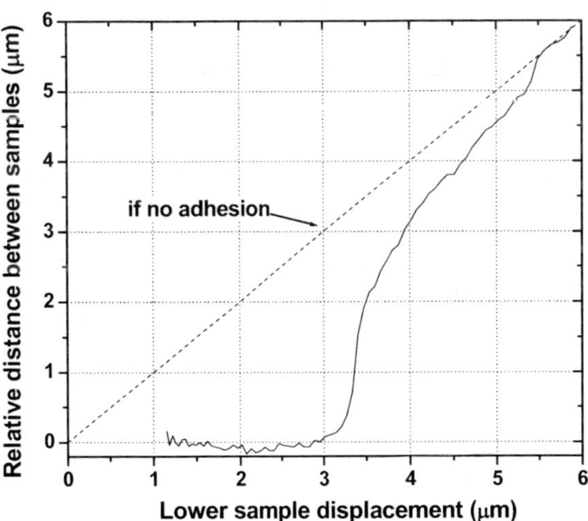

Fig. 3 Relative distance change between two sample surfaces during the PZT actuation as a function of the lower sample displacement from the calibrated zero position.

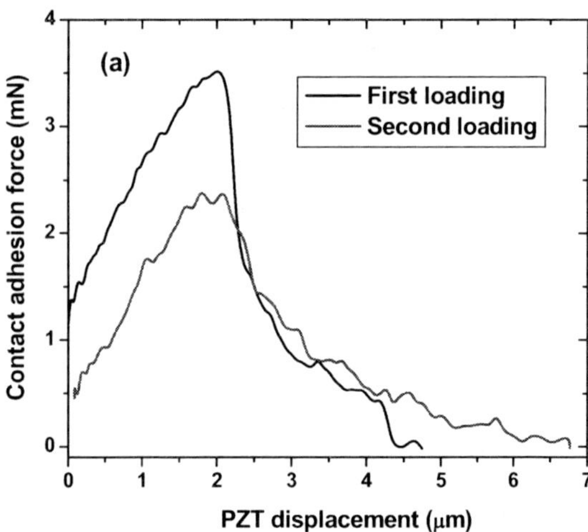

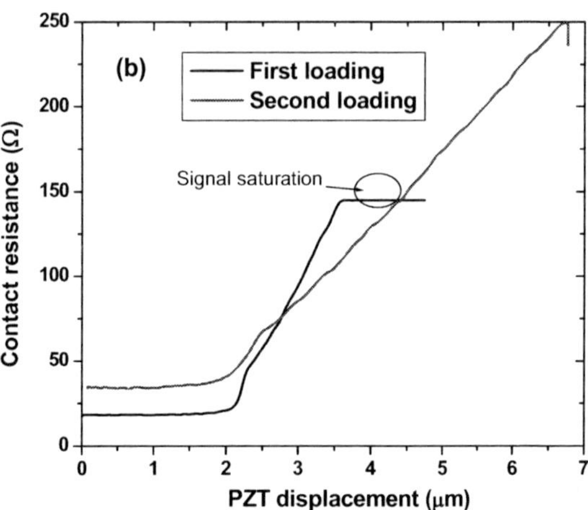

Fig. 4. (a) Adhesion force and (b) contact resistance measurement for MWNT-MWNF brush contact. With a contact area of ~8 mm^2, the maximum adhesion force realized in the experiment was about 3.5 mN and the contact resistance in the close contact region was about ~20 Ω for the first loading; for the second loading, the maximum adhesion force is ~2.3 mN and the minimum contact resistance is ~35 Ω.

$$k = 3\left(\frac{2\pi f_1}{3.6637}\right)^2 \rho abL \qquad (3)$$

With the two types of cantilevers used in this work, quartz cantilever of 0.5 mm thick and glass cantilever of 1 mm thick, varying the length of the cantilevers (30-50 mm) can give a spring constant ranging from 100 N/m to more than 1000 N/m.

The contact resistance is measured by four-terminal method as indicated in the experiment schematics. A constant current is fed through two electrodes across the interface, and the voltage drop across the interface is measured through the other two electrodes. The contact resistance can then be determined by the ratio between the voltage drop and the current.

RESULTS AND DISCUSSION

Adhesion and contact resistance between MWNT and MWNF arrays

The MWNT array has a curly surface structure with nanotube tips entangling each other (Fig. 2(a-c)), while the nanofibers are generally thicker and straighter. It is thus expected that upon engagement of the two surfaces, the stiffer and straighter nanofibers tend to stick into the curly surface of the nanotubes generating adhesion force between the two surfaces as a result of the mechanical interaction and van der Waals forces.

During the adhesion force measurement, the MWNF sample, which sits on the lower glass substrate, was manually brought up into contact with the MWNT sample at the free tip of the cantilever. An initial pressure force of ~10 mN was applied between the two samples for their engagement. The stage was then adjusted back slowly to relax the cantilever to near zero pressure while maintaining close contact between the samples. The lower sample was lowered further very slowly by actuating the PZT. The actuation speed was 2V/sec, ramping from 100 V to 0 V. The deflection of the cantilever was monitored from the reflected laser spot on the PSD during the process to determine the displacement of the upper nanotube sample.

Figure 3 shows the relative distance between the two samples during the PZT actuation. From the relative distance curve, a clear adhesion and separation trend can be identified. As the lower sample moved down, the upper sample first displaced with the lower sample due to adhesion. At ~3 μm of the displacement of the lower sample from zero position, the two samples started separation. They seemed to have separated gradually probably due to the tearing at the interface. With the spring constant of the cantilever, the adhesion or pressure force can be calculated. The contact resistance between the two samples was also monitored during the process by the four-terminal method discussed above. The contact adhesion force and the contact resistance as a function of the PZT displacement are shown in Fig. 4(a,b). As indicated in the graph, the maximum adhesion force between the two samples upon first loading is 3.5 mN and the minimum contact resistance is about 20 Ω. With the apparent contact area of 2 mm by 4 mm, the adhesion force per unit area is ~438 N/m^2, and the contact resistivity is ~1.6 $\Omega \cdot$cm^2. Considering an area density of 10^8 fibers/cm^2, the adhesion force per fiber contact is ~0.35 nN, about 10% of the theoretical prediction of the 'nano-hook' pair [14]. The contact force experiment is conducted a second time to test the repeatability of the process. The second loading gave maximum adhesion of 2.3 mN and minimum resistance ~35 Ω, showing the surfaces might degrade during contact.

The SEM pictures confirmed the surface degradation of the MWNF surface (Fig. 2(d-f)). While the carbon nanotube surface (Fig. 2(b)) did not show obvious changes, some of the carbon nanofibers were uprooted from their substrate and stuck into the nanotubes' curly surface. In Fig. 2(e), one can see blobs of metal underlayer on the top of uprooted nanofibers, which were originally at the nanofiber-substrate interface. And in Fig. 2(f) of the magnified view of the penetration of the nanofibers into the nanotube surface, one can identify that at the tip of each fiber there is a shiny particle, which is the catalyst particle responsible for the tip growth mechanism of the nanofibers. This agrees with general observations of the MWNT and MWNF samples in this work. The nanotube samples in this work are well attached to the substrate that one needs to use a tweezer with considerable force to scratch the nanotubes off the substrate, while for the nanofiber samples one can easily 'wipe' off the nanofibers. Thus to make reliable interfaces between the MWNT and MWNF arrays, the attachment of the nanofibers to the substrate is a concern and needs further investigation.

While the MWNTs are well attached to the substrate, the adhesion force between two such nanotube samples (i.e., MWNT-MWNT contact) is comparatively small, < 20 N/m^2. This may be due to the fact that the carbon nanotube bundles are long (~100 μm), thus less stiff, and the surfaces are relatively dense and curly such that they are harder to penetrating each other. As we have experimented with some shorter MWNT samples of ~8 μm tube length grown with shorter time (30 sec), the adhesion force increased to at least 100 N/m^2 with good repeatability.

<u>Contact resistance of MWNT-MWNT and MWNT-gold interfaces</u>

Contact resistances are measured for the MWNT-MWNT contact and MWNT-gold surface contact. The two samples are pressed against each other and the contact resistance is monitored as a function of the pressure force. Several groups of resistance data are plotted in Fig. 5. The square and round data series represent two MWNT-MWNT surface contact experiments; the up and down triangle data series represent two MWNT-gold surface contact experiments. The formed loops for the round series and the two triangle series are the resistances change while ramping up the pressure and then down. The minimum contact resistances for the MWNT-MWNT contact and MWNT-gold contact are ~20 Ω and for a contact area of 2 mm by 2 mm the contact resistivity is also on the order of 1 Ω·cm^2.

We have also varied the nanotube lengths to see whether there is any length dependence of the contact resistances for the MWNT-MWNT contacts. A group of samples with different nanotube lengths (~8 μm, ~20 μm, ~30 μm, and ~100 μm) were tested and the measured minimum

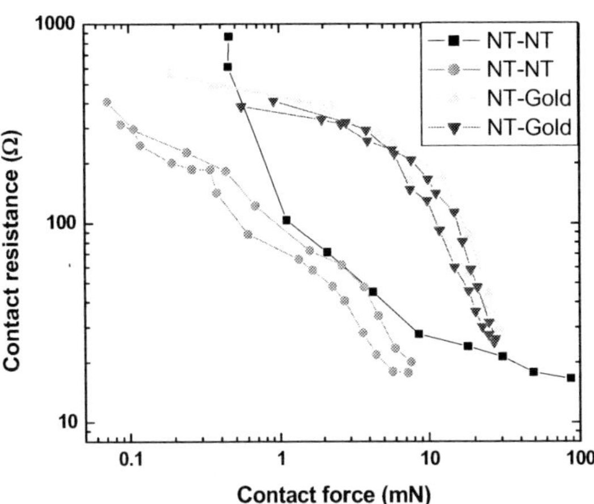

Fig. 5 Electrical contact resistance measurements of MWNT-MWNT surface contact and MWNT-gold surface contact. The square and round data points series represent two MWNT-MWNT surface contact experiments; the up and down triangle data points series represent two MWNT-gold surface contact experiments. The contact force is the pressure force between two samples. The formed loops for the round series and the two triangle series are the resistance change during ramping up the pressure force and then down.

contact resistances range from ~1 Ω to ~100 Ω with no clear dependence on the tube lengths. This shows either the electronic transport through the MWNTs is still ballistic at this length scale or the length dependent resistance is small compared with nanotube-nanotube and/or nanotube-substrate interface resistances. In this work, resistances of nanotube-nanotube contact interface and nanotube-substrate interface cannot be determined independently. Even for the smallest measured contact resistance, the per-tube contact resistance is huge, > 100 MΩ. There might be several reasons for this: first, at the nanotube-nanotube contact surface, the nanotubes are mainly side-contacted, and the cross-tube contact resistance might be very large; second, the inter-penetration is limited due to the high density of the nanotubes. Further investigations are needed to study the contact characteristics at the interface.

CONCLUSION

Multiwalled carbon nanotubes and nanofibers grown by CVD method have the potential of being used as conductive and dry adhesive interface materials for electrical contacts. We demonstrate in this work an adhesion force between MWNT and MWNF array surfaces of 3.5 mN for an apparent contact area of 2mm by 4 mm, thus 438 N/m^2 for adhesion per unit area. Though higher than the demonstrated adhesion between short MWNT surfaces (~ 8 μm tube lengths) of 100 N/m^2, MWNF surfaces in this work degraded with contacts while

MWNT surfaces did not. The minimum contact resistance measured at the same time is ~20 Ω. Contact resistances of MWNT-MWNT and MWNT-gold interfaces were also measured as pressure forces around several mN were applied at the interface. The resulting minimum contact resistances are on the same order but with considerable variation from sample to sample. For MWNT-MWNT contacts, a minimum contact resistance of ~ 1 Ω is observed for a contact area of 2 mm by 1 mm. The relatively high contact resistances per-tube, considering the area density of the nanotubes, might be explained by the high cross-tube resistances at the contact interfaces or limited inter-penetration due to the high density of the nanotubes array. Further investigations are needed to explain the contact characteristics at the interface.

Carbon nanotube interfaces acting as thermal switches are also of tremendous interest. They may play an important role in the cryogenic devices and are thus our future research objective.

ACKNOWLEDGMENTS

The authors gratefully acknowledge the financial support for this work from NASA SBIR program. The authors also thank the Berkeley Microlab for the fabrication of part of the experiment devices.

REFERENCES

[1]   D. Hyman and M. Mehregany, Contact Physics of Gold Microcontacts for MEMS Switches, IEEE Transactions on Components and Packaging Technologies, 22 (1999) 357-364.

[2]   E.J.J. Kruglick and K.S.J. Pister, Lateral MEMS Microcontact Considerations, Journal of Microelectromechanical Systems, 8 (1999) 264-271.

[3]   L.L. Mercado, T.Y.T. Lee, S.M. Kuo, V. Hause, and C. Amrine, Thermal Solutions for Discrete and Wafer-Level RF MEMS Switch Packages, IEEE Transactions on Advanced Packaging, 26 (2003) 318-326.

[4]   Forro, L. and Schonenberger, C. Physical Properties of Multi-Wall Nanotubes.; 80 pp. 329-390.

[5]   L. Delzeit, I. Mcaninch, B.A. Cruden, D. Hash, B. Chen, J. Han, and M. Meyyappan, Growth of Multiwall Carbon Nanotubes in an Inductively Coupled Plasma Reactor, Journal of Applied Physics, 91 (2002) 6027-6033.

[6]   L. Shi, PhD Dissertation, Mechanical Engineering Department, UC Berkeley, Http://Nano.Me.Berkeley.Edu/Dissertations/Lishi_Phd_Dissertation.Pdf, (2001) .

[7]   L. Delzeit, C.V. Nguyen, B. Chen, R. Stevens, A. Cassell, J. Han, and M. Meyyappan, Multiwalled Carbon Nanotubes by Chemical Vapor Deposition Using Multilayered Metal Catalysts, Journal of Physical Chemistry B, 106 (2002) 5629-5635.

[8]   C.T. White and T.N. Todorov, Carbon Nanotubes as Long Ballistic Conductors, Nature, 393 (1998) 240-242.

[9]   H.T. Soh, C.F. Quate, A.F. Morpurgo, C.M. Marcus, J. Kong, and H.J. Dai, Integrated Nanotube Circuits: Controlled Growth and Ohmic Contacting of Single-Walled Carbon Nanotubes, Applied Physics Letters, 75 (1999) 627-629.

[10] P.J. De Pablo, E. Graugnard, B. Walsh, R.P. Andres, S. Datta, and R. Reifenberger, A Simple, Reliable Technique for Making Electrical Contact to Multiwalled Carbon Nanotubes, Applied Physics Letters, 74 (1999) 323-325.

[11] S. Frank, P. Poncharal, Z.L. Wang, and W.A. De Heer, Carbon Nanotube Quantum Resistors, Science, 280 (1998) 1744-1746.

[12] S.K. Biswas, R. Vajtai, B.Q. Wei, G.W. Meng, L.J. Schowalter, and P.M. Ajayan, Vertically Aligned Conductive Carbon Nanotube Junctions and Arrays for Device Applications, Applied Physics Letters , 84 (2004) 2889-2891.

[13] J. Li, R. Stevens, L. Delzeit, H.T. Ng, A. Cassell, J. Han, and M. Meyyappan, Electronic Properties of Multiwalled Carbon Nanotubes in an Embedded Vertical Array, Applied Physics Letters, 81 (2002) 910-912.

[14] S. Berber, Y.K. Kwon, and D. Tomanek, Bonding and Energy Dissipation in a Nanohook Assembly, Physical Review Letters, 91 (2003).

[15] O. Englander, D. Christensen, and L.W. Lin, Local Synthesis of Silicon Nanowires and Carbon Nanotubes on Microbridges, Applied Physics Letters, 82 (2003) 4797-4799.

[16] Vibrationdata.com, http://www.vibrationdata.com/tutorials/plate.pdf.

NANO2004-46016

RECONSTITUTING MEMBRANE PROTEINS INTO ARTIFICIAL MEMBRANES AND DETECTION OF THEIR ACTIVITIES

Hyeseung Lee, Dean Ho, Benjamin Chu, Karen Kuo, Carlo Montemagno

Department of Bioengineering, University of California at Los Angeles, CA 90095

ABSTRACT

We have successfully purified BR from purple membrane of *Halobacterium Salinarium* and Cox from the genetically engineered plasmid inserted in *Rhodobacter Sphaeroides*. The activities of the purified enzymes have shown in lipid vesicles as well as in polymer vesicles and planar membranes. Phosphatidylcholine derived lipid vesicles created the most nature like environment for the enzymes. Triblock copolymer membrane was the alternative choice for membrane protein reconstitution since polymers are more durable, ideal for industrial applications and support enzyme activities better. We also demonstrated the backward function of Cox *in vitro* by changing proton concentration in the surrounding medium. Langmuir-Blodgett method was used to reconstitute the enzymes into the planar lipid or polymer membranes. The enzyme activities of the enzymes in planar membrane system were tested by impedance spectroscopy.

INTRODUCTION

With the evident limitations in size miniaturization in conventional batteries, there is a need to develop an alternative power supply to drive more compact devices in future technology. As a potential nano-scale hybrid device which generates energy from biological source, the biocompatible fabric embedded with energy converting proteins was explored. There are two different types of energy converting proteins embedded in artificial membrane: bacteriorhodopsin (BR) and cytochrome oxidase (COX). Upon the presence of light, bacteriorhodopsin starts pumping protons from one side to another side of the membrane creating proton gradient. This electro-chemical gradient across the membrane forces cytochrome oxidase to function in reversed action. Reversed Cox mechanism generates intermediates from O_2 to H_2O along with electrons. The electrons are detected and attracted to the electrode placed in the vicinity of the protein in the membrane.

The system converts optical energy to electrical energy, eventually allocating the derived energy to an external source.

Bacteriorhodopsin is one of the most widely studied ion transport proteins whose activity as a proton transporter has been conserved by natural selection over millions of years. Upon the presence of green light (λ=500-650nm). BR molecules in purple membrane perform unidirectional pumping of protons from the cytoplasm to the extracellular space during the photocycle. An electrochemical gradient across the membrane is formed as each BR molecules undergo various electronic intermediate states, with the total time of less than 3ms.

Cytochrome oxidase is the terminal enzyme of the respiratory chain in mitochondria. Cytochrome c sits at the specific docking site exposed in the external side of Cox to donate electrons. The electrons are translocated in cascade within the protein, finally reducing O_2 to H_2O at the lumenal side. The potential drop (460mV) created by electron transfer is used for vectorial translocation of protons to generate a transmembrane pH gradient. Studies have shown that it is possible to reverse COX function partially by adding ATP to mitochondria [1, 9]. In the presence of ATP, ATPase in mitochondria pumps protons out creating pH gradient across the membrane. The eletro-osmotic proton pressure was formed by the high proton concentration on the external side of mitochondria. This pressure drives COX backwards, resulting in electron transfers from water to the protein.

MATERIALS AND METHOD

A. Bacterial culture and protein purification

Bacteriorhodopsin was isolated from the purple membrane sheets purchased from COBEL. BR was purified as described in Dencher et al [2]. Cytochrome c oxidase over-expression system was kindly donated by Dr. Ferguson-Miller at Michigan

State University. The bacterial culture conditions and purification of Cox were explained by Zhen et al [3].

B. Lipid vesicle formation and protein activity detection in vesicles

Liposomes and proteoliposomes were prepared mainly as described in Hazard et al [4]. The enzyme concentrations in proteoliposomes were 90μg/mL for COX and 137μg/mL for BR. Fluorescent probe, pyranine was trapped within the vesicles to indicate the pH change by the proteins. Perkin Elmer Fluorometer was used to monitor the fluorescent. The oxygen depletion in Cox proteoliposomes was measured by WTW Oxi-Cal. Spectrohotometry was performed by Beckmann UV-Visible spectrophotometer. The substrate for COX was horse heart cytochrome C (sigma) pre-reduced by dithionite and desalted through Sephadex G-25 (Amersham). BR was activated by a green-light filter enhanced 150W Fostec light source.

C. Polymer vesicle formation

Triblock polymer (PMOXA-PDMS-PMOXA) was dissolved in ethanol to 17wt%. Bacteriorhodopsin/Block Copolymer hybrid vesicles were formed by adding 5mg/ml protein very slowly to this polymer solution at 4°C. Then, pyranine was introduced to this solution in the same manner and the mixture was dialyzed at 4°C overnight. The vesicles were syringe-filtered through 200nm filters in order to develop polymer vesicles of 250nm diameter. Cross-linking of the polymer was performed by UVP 400W 254nm Spotcure system. Irradiation was conducted at a distance of 5mm from the cuvette for 15 minutes.

D. Impedance spectroscopy in planar polymer membrane

This is a three electrode system where Ag/AgCl electrode is the reference electrode and platinum electrode serves as counter electrode. The gold film evaporated on glass serves not only as a solid support for the membrane deposition but also as a working electrode. Gold is pre-evaporated in 150nm thickness on glass or quartz. Monolayer polymer membrane and enzymes are deposited by Langmuir-Blodgett method. The membrane deposited substrate is placed at the bottom of the Teflon chamber body where the opening (radius=1mm) of the cone shape inner space is. This space is filled with buffer (0.1M KCl, 0.01M tricine, 0.01M Na_2HPO_4, 0.0002M $MgSO_4$, pH 7.3) and any activating/deactivating substrate is added in here. The frequency was varied from 100kHz to 0.1Hz with the constant amplitude of 10mV.

RESULTS

A. Protein isolations

The two strains of *Rhodobacter Sphaeroides* (YZ100 and 37-2) carried genetically engineered plasmid for Cox. Nickel NTA is used to isolate Cox from the solubilized membrane since it carried six histidine residues which is easily attracted to metals. In some cases, the metal affinity chromatography was followed by anion exchange chromatography (FPLC) to achieve higher purity. The recovery of Cox was less than 10% from FPLC and the loss in functionality was seen from the oxidase purified by FPLC. Nickel NTA gave enough purity in Cox for the most activity assays. When estimating COX amounts, the extinction coefficient of $\Delta\varepsilon\,^{COX}_{606-630}$ =28.9mM^{-1} cm^{-1} was used for dithionite-reduced minus ferricyanide-oxidized difference spectra by UV-Visible spectrophotometer. All steps were done at 4 C° to prevent denaturing of membrane proteins. BR purification was done in dark to minimize the unwanted proton pumping activity. BR concentration was estimated by the extinction coefficient of $\Delta\varepsilon\,^{BR}_{568}$ =63mM^{-1} cm^{-1}.

B. Monitoring internal pH changes of proteoliposomes
Phosphatidylcholine and phosphatidyl acid were first dissolved in ether (Hazard). Using rotary evaporator ether in the mixture was slowly removed and larger vesicles formed. Subsequent syringe filtrations (pore size: 0.45μm, 0.20μm) eliminated multilamellar or oligolamellar vesicles. By introducing cholesterol into the lipid mixture, the vesicles became more impermeable to protons, which gave accurate measurement of proton transport by proteins. The previous studies have shown that the radius of lipid vesicles prepared in similar manner was 15nm, with 5-20 molecules of oxidase per vesicle [5]. Our vesicle size ranged from 200nm to 10nm. Once dissociated from the membranes, the membrane bound proteins became unstable and started degradation. When incorporated into the artificial membranes, proteins gained their structure, therefore, functionality. The proteins maintained their activity up to a month when stored at 4 C° in phosphatidylcholine based lipid vesicles.

Pyranine is a fluorescent probe whose fluorescence intensity varies depending on the pH. Proteoliposomes (lipid vesicles) were formed in the presence of pyranine for measurement of changes in internal pH. External pyranine was eliminated by dialysis, and the fluorescence of remaining external pyranine was quenched by the addition of *p*-xylenebispyridinium bromide (DPX). DPX is membrane impermeable, therefore, only external pyranine is removed, leaving only the fluorescence from the internal pyranine detectable. An electrochemical gradient ($\Delta\mu H^+$) is produced by the movement of protons across the membrane. Both electrical ($\Delta\psi$) and chemical (ΔpH) components of the gradient produce "back pressure effects" which limit maximum attainable change in $\Delta\mu H^+$. By eliminating the electrical component of this backpressure, it is possible to increase ΔpH. This can be accomplished using the K^+ ionophore valinomycin, which neutralizes the charge across the membrane [4], leaving only the H^+ gradient. Addition of Valinomycin increased the maximum pH gradient, but did not improve the enzyme activity itself. This pH gradient formation was inhibited and eventually reversed by the uncoupler CCCP.

c. COX activity in lipid vesicles

In the forward reaction of Cox, electrons are donated from cytochrome c to cytochrome oxidase, eventually in the lumenal side those eletrons are used to reduce oxygen molecules into water with the ambient protons ($4H^+ + O_2 \rightarrow 2H_2O$). At the same time, there is a vectorial translocation of hydrogen to the external side. The same reactions were shown *in vitro* in Cox proteoliposomes when pre-reduced cytochrome c was added. Upto the saturation point, the linear relationship was observed between pH and the amount of cyt C added (figure 1). Since the protons were transported out of the vesicles, increase in pH was observed within the vesicles. Oxygmeter monitored the oxygen depletion of COX proteoliposomes (figure 2). The constant relflux of oxygen at the surface of proteoliposome solution was taken account to estimate the oxygen depletion in Cox proteoliposomes. The total of 2.3mg/L of O_2 depletion occurred in 60s (0.12nmol/sec). A total of 22.5µg of COX was embedded in lipid vesicles of the experiment. Since no attempt to orient the proteins was made at this point, it was assumed that only half of COX ($9*10^{-2}$ nmol) was correctly oriented, exposing the docking site of subunit II available for cyt C. The turnover number was estimated approximately as 133/sec, which was higher than the reported values of 95/sec or 106/sec [6].

For Cox reversal assays, extra protons were introduced to the medium by adding 0.07M HCl dropwise. The outside pH was measured by pH probe directly. The internal pH of the vesicles was monitored by pyranine entrapped inside vesicles under the fluorometer. At the initial stage of the experiment, the sufficient amount of cytochrome c was introduced to the COX proteoliposome solution to completely reduce the heme centers of COX. At this point, Cox proteoliposomes were depleted in protons due to the forward activity of Cox. Proton depleted inner vesicle space was the perfect condition to activate COX reversal with lower potential barrier to overcome. Any excess proton outside the vesicle rushed through the enzymes reversing the electron pathway of COX. The phenomenon was described as follows [7]: $E_2-E_1=\Delta G/nF$, where E_2 and E_1 are the respective redox potentials on the high and low potential sides of the energy-conservation site, ΔG is the Gibbs free energy change, n is the number of electrons transferred, and F is the Faraday constant. As the external pH increased, ΔG was enhanced, favoring reversed electron transfer. As the electrons travel through the protein, the heme centers of metal groups undergo conformational changes depending on the stage of the electron transfer. This can be detected by optical spectrophotometry, by which oxido-reduction of the hemes was monitored (figure 3). The total of $2.32*10^{17}$ numbers of hydrogen ions were added in 150µL of proteoliposome solution ($\Delta pH = 2.59$, $\Delta pH = -\log [H^+]$). In forward reaction, 4 protons are transported upon the consumption of 4 electrons. If all the protons added from HCl were used to drive COX backwards, it

can be said that the total of 0.0372C of electrical charge ($q=1.602*10^{-19}$ C) was generated by $1.1*10^{13}$ numbers of COX in its reversed activity.

D. BR activity in lipid vesicles & polymer vesicles

The 150µL of BR proteoliposomes in a cuvette (enzyme concentration of 130µg/mL) formed a light-driven pH gradient of 0.6 unit in less than 1 minute in the absence of valinomycin (figure 4). Rapid pH change was observed at the initial stage of light illumination. It was reported that the ΔpH formed in the presence of 0.1µM valinomycin was 70% lower in proteoliposomes of BR, whether valinomycin was added before or after the onset of proton pumping. They attributed the inhibitory effect of valinomycin on the ΔpH formation to the direct interaction of this antibiotic with bacteriorhodopsin, which has been demonstrated to interfere with the enzyme's photocycle and proton pumping activity [8]. The pH change measured in the absence of valinomycin was considered more reflective of the experimental conditions which would be encountered in more advanced system where cytochrome c oxidase and bacteriorhodopsin exist at the same time and one activates the other. We did not attempt to orient BR in this experiment. It is possible that the enzymes oriented in opposite direction end up canceling the ΔpH by pumping the protons in the opposite side of the lipid vesicle. If there is a cross orientation of equal numbers of participating BR molecules, the cooperative effect disappears.

In BR/Polymer system, a rapid pH change could be seen upon illumination of the light (figure 5). Subsequent readings, however, indicated a rapid decrease in pH change, until negligible change was observed. With non-crosslinked vesicles, a maximum change of 0.17 was seen in the pH while crosslinked vesicles produced a maximum pH change of 0.34. In both cases, however, there was a strong indication that while initial protein activity was the likely mechanism for the rapid pH changes, subsequent tapering of pH activity was due to the permeability of the polymer vesicles to proton flow, thus precluding formation of a lasting gradient for protein activity. Such indications are supported by the fact that the block copolymer retains a rigid, rod-like behavior which reduces membrane continuity and fluidity found in lipid bilayer systems which are paramount to the formation of impermeable walls. Using shortwave crosslinking of the highly-reactive acrylate endgroups, however, results showed the increase in pH change, probably due to reduction in membrane permeability due to endgroup reactions that created a more effective seal in the polymer. Leakage of pyranine through the polymer walls did not seem to contribute to the low pH change values as initial pH readings of 7.3 correspond to the pH of the unreacted pyranine, indicating that the pyranine was indeed trapped inside the vesicle. Subsequent work will concern the implementation of secondary sealants to improve membrane resistance to proton flux.

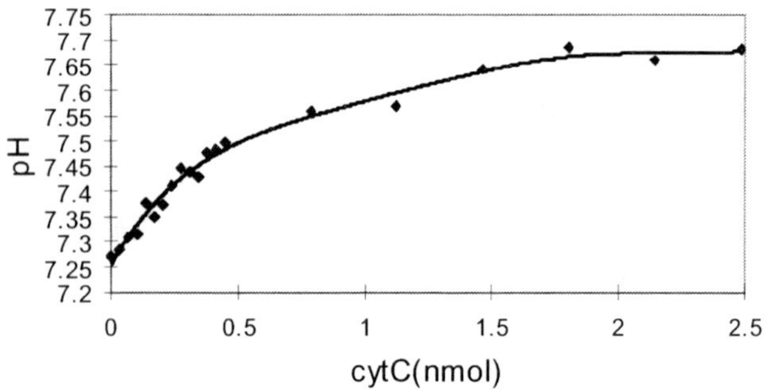

Figure 1. Forward Cox functionality in proteoliposomes. 0.034nmol cyt C added first 13 times, then 0.34nmol cyt C added every minute into 13.5µg COX reconstituted in proteoliposomes. Upon the addition of cytochrome c, cytochrome c oxidase started pumping protons out of the vesicles, increasing the pH of the internal side of the membrane. Notice that the first 14 data points made a linearly increasing slope while the later 6 data points had a smaller slope. The rate of formation of steady-state ΔpH was greater at the earlier stage of cytochrome c addition.

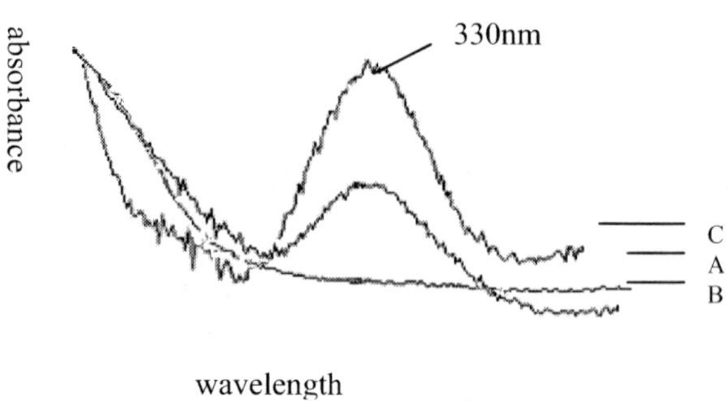

Figure 3. Reversal of COX monitored by spectrophotometer. **Bold: external pH**, *italic: internal pH*; a. **7.39**, *9.5*, b. **5.45**, *7.4*, c. **4.8**, *7.3*. Notice formation of the 330nm absorption peak of heme a. The spectral shift that we saw was clearly the consequence of energy-linked reversed electron transfer. Although O_2 generation from water has not been demonstrated despite numerous attempts, a partial reversal of the O_2 reaction remains possible [1].

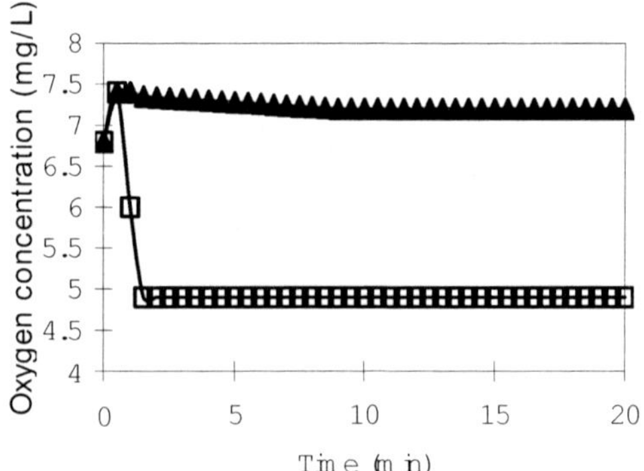

Figure 2. Oxygen depletion in COX proteoliposomes. 400µL of proteoliposome solution (90µg/mL COX concentration) was brought up to 10mL with lipid dialysis buffer. 60nmol cyt C added at time zero. Initial oxygen content increase occurred due to the stir mixing. Triangles: control of no COX lipid vesicles. Squares: COX proteoliposomes.

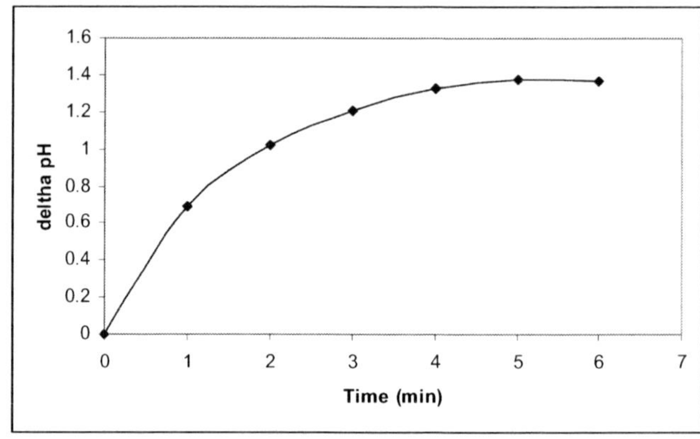

Figure 4. proton pumping activity of BR in proteoliposomes. ΔpH is plotted according to time laps. At time 0, green light illumination started.

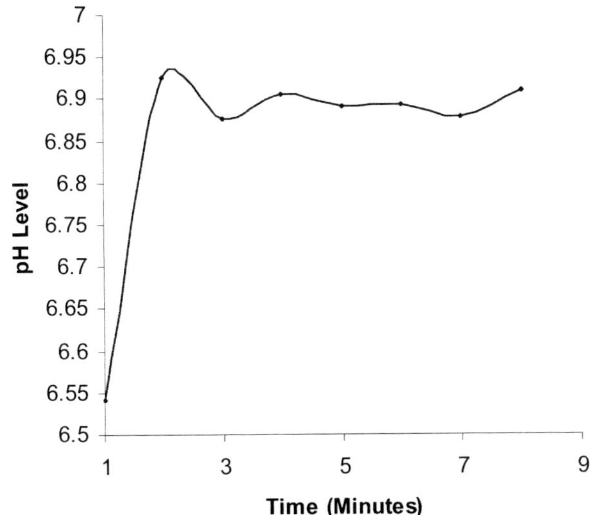

Figure 5. proton pumping activity of BR in polymer vesicles. Since BR pumps proton out of the vesicles, pH increase observed at the initial stage. Constant reflux of pH was seen due to the rigid nature of the polymer membrane.

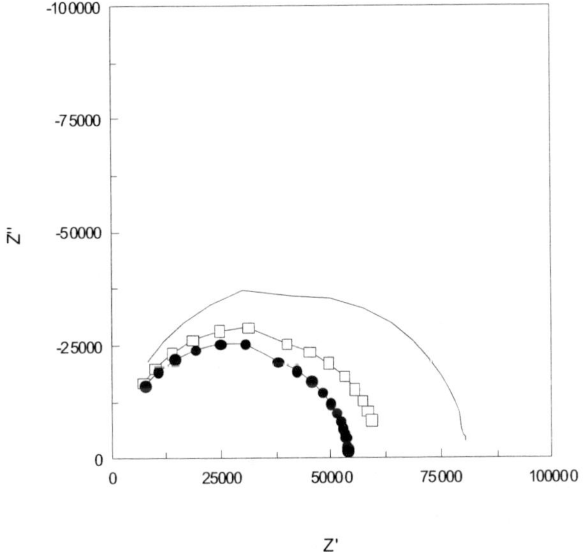

Figure 6. Impedance spectroscopy of polymer monolayer with BR inserted by LB method. The membrane and protein was deposited on quartz or silicon dioxide substrate evaporated with gold on top. The solid line: initial point (no light), open square: light illumination (1min), dark circle: light illumination (2min).

E. Planar membrane formation

Langmuir-Blodgett technique is a technique fabricating ultra-thin anisotropic layers from molecules of an organic/inorganic substance. Most molecules forming satisfactory monolayers have an asymmetric structure. A Langmuir monolayer is an individual monolayer made of amphiphilic molecules which organize themselves at the contact with an air-water interface. Lipids, membrane proteins and triblock copolymer in this study are good examples of amphiphiles applicable in LB deposition. Proteins were injected to the LB trough after the monolayer formation on the water surface and the incubation time of 45min was given to insure the protein incorporation to the membranes. Most depositions were performed at 40mN/m or higher. Due to the rigidity of the polymer membrane, surface pressure higher than 50mN/m was seen to break the langmuir monolayer.

F. protein activity detection in planar membrane

Impedance spectra were recorded to verify the presence of monolayer membrane and enzyme activity detection in the planar membrane system. The semi-circle at high frequencies is related to the parallel combination of the double-layer capacitance and the charge transfer resistance from the polymer monolayer. At high frequencies (fast scan rate), impedance is determined by slow electron transfer kinetics through the polymer membrane. Since the electron transport through the enzyme is faster than nonfaradaic charge transfer in the polymer membrane, the decrease in impedance occurs in the presence of protein activity (figure 6). Upon the light illumination, protons are pumped through the BR molecules in polymer membrane decreasing the insulation behavior of the thin layer. Several studies have reported impedance behavior from biomimeric membranes. When the membrane was deposited by fusion of vesicles method, discrepancy was reported which is assumed to be due to the packed vesicle aggregates and pinholes of gaps in the membrane [10]. Langmuir-Blodgett deposition gave uniform planar monolayers with surface roughness of the substrate being the biggest challenge.

SUMMARY AND ONGOING WORK

The two membrane proteins of interest, BR and COX were successfully purified from the cell membranes of *Halobacterium Salinarium* and *Rhodobacter Sphaeroides* respectively. The proton transporting activity of COX and BR has been demonstrated in lipid vesicles made of phosphatidylcholine and cholesterol. It was shown that it is possible to drive COX backward by varying proton concentrations. BR activity was also shown in polymer vesicles. Several attempts were made to orient proteins in desired direction for the maximal device function. The cystein residue exposed in the external side in BR is the linker between the protein to the gold surface on the substrate. Cox can be oriented in either direction depending on the location of the histidine tags on the protein. YZ100 produces subunit I his tagged enzymes whose tail is exposed in the luminal side while 37-2 makes subunit II his tagged enzymes which is located in the external side of the membrane. To make the device more stable and applicable to industrial purposes, the device assembly is cast into tetramethyl orthosilicate based sol gel $(Si(OCH_3)_4)$. Direct current measurement from the combine

17

protein hybrid device is conducted by cyclic voltammetry (data not shown).

ACKNOWLEDGMENTS

We thank Dr. Ferguson-Miller group at Michigan state university for providing us with Cox over-expression system of two strains of *Rhodobacter Sphaeroides*: YZ100 and 37-2.

This research is funded by DARPA.

REFERENCES

1. M. Wikstrom, and J. E. Morgan, "The dioxygen-cycle," The journal of biological chemistry, vol. 267, pp. 10266-10273, 1992.

2. N. A. Dencher and M. P. Heyn, "Preparation and properties of monomeric bacteriorhodopsin," Methods in enzymology, vol. 88, pp. 5-10, 1982.

3. Y. Zhen, J. Qian, K. Follmann, T. Hayward, T. Nilsson, M. Dahn, Y. Hilmi, A. G. Hamer, J. P. Hosler, and S. Ferguson-Miller, "Overexpression and purification of cytochrome c oxidase from Rhodobacter sphaeroides," Protein expression and purification, vol. 13, pp. 326-336, 1998.

4. A. Hazard, C. Montemagno, "Improved purification for thermophilic $F_1 F_0$ ATP synthase using *n*-dodecyl ß-D-maltoside," Archives of biochemistry and biophysics, vol. 407, pp. 117-124, 2002.

5. C. Hiser, D. Mills, M. Schall, S. Ferguson-Miller, "C-terminal truncation and histidine tagging of cytochrome c oxidase subunit II reveals the native processing site, shows involvement of the c-terminus in cytochrome c binding, and improves the assay for proton pumping," Biochemistry, vol. 40, pp. 1606-1615, 2001.

6. Y. A. Ivashchuk-Kienbaum, "Monitoring of the membrane potential in proteoliposomes with incorporated COX using the fluorescent dye indocyanine," J. Membrane Biol., vol. 151, pp. 247-259, 1996.

7. D. De Vault, biochim. Biophys. Acta, vol. 226, pp. 193-199, 1971.

8. C. A. Hasselbacher, T. G. Dewey, "Characterization of the binding of Valinomycin to bacteriorhodopsin," Photosynth. Res., vol. 8, pp. 79-86, 1986.

9. M. Wikstrom, "Energy-dependent reversal of the cytochrome oxidase reaction," Proc. Natl. Acad. Sci. USA, vol 78, pp. 4051-4054, July 1981.

10. T. Baumgart, M. Kreiter, H. Lauer, R. Naumann, G. Jung, A. Jonczyk, A. Offenhausser, W. Knoll, "Fusion of small unilamellar vesicles onto laterally mixed self-assembled monolayers of thiolipopeptides," Journal of colloid and interface sciences, vol 258, pp. 298-309, 2003.

NANO2004-46018

PROTEIN-FUNCTIONALIZED PROTON EXCHANGE MEMBRANES

BENJAMIN CHU, DEAN HO, HYESEUNG LEE, KAREN KUO, AND CARLO MONTEMAGNO

Bioengineering Department
University of California at Los Angeles, Los Angeles, CA 90095

ABSTRACT

Protein-functionalized biomimetic membranes, based upon a triblock copolymer simulating a natural lipid bilayer in a single chain, serves as a core technology for applications in bioenergetics. Monolayers of block copolymer, which simulates the hydrophilic-hydrophobic-hydrophilic chain of a natural cell membrane, can be formed by Langmuir-Blodgett (LB) deposition and provides a favorable environment for protein refolding. Large-scale membrane formation is achieved using LB deposition on a variety of substrates, such as gold, quartz, silicon, and Nafion®. We have successfully inserted membrane proteins, such as the light-activated proton pump, bacteriorhodopsin (BR) and the pH/voltage-gateable porin, Outermembrane Protein F (OmpF), into large-area LB monolayers. We have also established sustained protein functionality in films through the measurement of light-activated proton transport.

INTRODUCTION

Protein-functionalized biomimetic membranes integrated with proton exchange membranes can serve to form a more efficient fuel cell. Conventional fuel cell technology, while producing a pollution-free high-density energy source, continue to suffer from inefficiencies due to proton leakage across proton exchange membranes such as Nafion®, a perflourosulfonic acid polymer consisting of a perfluorinated carbon backbone similar in structure to Teflon® with sulfonic acid side chains. The issue of proton-leakage can be addressed by integrating Nafion® with a protein-functionalized biomimetic membrane. Our core technology consists of a triblock copolymer, which mimics a natural lipid bilayer in a single ABA chain, providing a hydrophilic-hydrophobic-hydrophilic environment for the proper refolding of protein. Due to its amphiphilic properties, triblock copolymers can be spread across an air-water interface for monolayer formation using a Langmuir-Blodgett (LB) trough, and with UV-crosslinkable methacrylate endgroups, the triblock copolymer exhibits an increased robustness over conventional lipid bilayer systems. Large-scale membrane formation is achieved using LB deposition on a variety of substrates, such as gold, quartz, silicon, and Nafion®. We have demonstrated the ability to insert a variety of membrane proteins, such as the light-activated proton pump, bacteriorhodopsin (BR) and the pH/voltage-gateable porin, outermembrane protein F (OmpF), into large-area LB monolayers. The energy-transducing protein BR, found in the bacteria *Halobacterium Halobium*, is a light-actuated proton pump capable of enduring temperatures as high as 180° C, within the operating temperature of a fuel cell. Using 10 nm-size colloidal gold particles bound to OmpF with a genetically modified cysteine residue, we have confirmed, through TEM imaging, protein insertion into block copolymer LB films. We have also confirmed, through UV-vis spectrophotometry, the presence of BR on Nafion® sheets integrated with a purple membrane monolayer by Langmuir-Blodgett deposition. Protein functionality in hybrid protein-Nafion® constructs was maintained and measurable pH changes were detected in the presence of light, indicating proton-pumping activity from BR. Due to the sulfonic acid side chains, Nafion®, when submersed in water or salt buffers, readily donates protons to its surrounding media, resulting in significant pH changes in the media. Using purple membrane fragments of unpurified BR, we have shown the ability to slow the pH drop of the buffer in which our functionalized-Nafion® was placed, indicating a retention of protons due to light-activated proton pumping.

NOMENCLATURE

BR – Bacteriorhodopsin

FPLC – Fast Performance Liquid Chromatography
LB – Langmuir-Blodgett
OmpF – Outermembrane Protein F
PEM – Proton Exchange Membrane
PM – Purple Membrane
TEM – Transmission Electron Microscopy

MATERIALS AND METHODS

<u>Triblock copolymer preparation</u>

Triblock copolymer, composed of polymethyloxazoline-polydimethyl siloxane-polydimethyloxazoline (PMOXA-PDMS-PMOXA), was purchased from and custom synthesized by Polymer Source Inc. (Dorval, QC, Canada). The ABA (hydrophilic-hydrophobic-hydrophilic) triblock copolymer was 8.4 nm in thickness and terminated with UV-polymerizable methacrylate end groups. Membrane forming solutions were prepared by dissolving the polymer in chloroform at a concentration of 10mg/mL, in an amber vial to prevent UV-cross-linking of the methacrylate end groups. Polymer solution was then stirred overnight for 24 hours on a platform stirrer at 320 rpm to facilitate solubilization.

<u>Protein Purification</u>

A plasmid containing the OmpF protein was inserted into the BL21/DE3 E. coli expression system. The protein was extracted by breaking the cells in a homogenizer and spun down. The OmpF protein was then purified using both isoelectric focusing with FPLC (Amersham Pharmacia) as well as gel filtration. Engineering of the OmpF was conducted by substituting a cysteine residue at the Glu 183 site [1] via site-directed mutagenesis to facilitate bioconjugation with colloidal gold beads with the cysteine residue.

Unpurified bacteriorhodopsin was purchased from Sigma-Aldrich in the form of Purple Membrane (PM) fragments from *Halobacterium Halobium*. PM fragments were solubilized in amber vials, to reduce exposure to ambient light, in a solution of phosphate buffer and Triton X-100 (Sigma). 2 minutes of vortexing the solubilized PM was followed by ultrasonication for 2 minutes, then the sample was left to solubilize overnight for 24 hours. Samples were then syringe-filtered prior to purification in the FPLC column. The gel filtration size-exclustion chromatography FPLC column was pre-equilibrated with the same PO_4/Triton X-100 buffer used for protein solubilization and wrapped in aluminum foil to shield the protein from ambient light during the course of the purification. After protein purification, dialysis was performed to remove any remaining detergent in order to facilitate protein insertion at the air/water interface during LB deposition.

<u>Omp-F Preparation and Visualization</u>

LB experiments were performed using a Kibron Microtrough platform with platinum rod surface pressure measurement. Langmuir troughs were cleaned by an alternating, triple rinse of ethanol and Milli-Q purified water. 200 mesh nickel Transmission Electron Microscopy (TEM) grids with a Fomvar/carbon support film coating, purchased from Structure Probe, Inc./SPI Supplies (West Chester, PA), were attached via double stick tape to a diced silicon wafer (to provide structural support) and lowered into the subphase. A 10mg/ml triblock copolymer solution dissolved in chloroform was spread across the subphase (Milli-Q water) to a starting surface pressure of 15mN/m and chloroform was allowed to evaporate off of the air-water interface for ten minutes. For OmpF sample preparation, supernatant from the protein purification was injected into the subphase to a starting pressure of 25mN/m and allowed 2 hours to diffuse to the surface. Compression was performed in constant pressure mode at a rate of 20Å^2/chain/min to a pressure of 35 mN/m. The TEM grids were lifted at a rate of 1mm/min using the perpendicular dipping method [2].

TEM grids coated cysteine-tagged OmpF/polymer films were then incubated and washed twice with a blocking buffer to prevent nonspecific binding. The samples were then incubated for 10 minutes with Nanogold® particles (Nanoprobes, Yaphank, NY) to allow for direct interaction between gold and the cysteine residue. The conjugated sample was then washed twice more with blocking buffer, twice in PBS buffer, and then with de-gased Milli-Q water. Samples were dried and imaged under TEM (JEOL 100CX) at 200KV.

<u>BR-Nafion® Sample Preparation</u>

For initial LB-deposited trials, purple membrane was solubilized at 2mg/mL in dimethylformamide (DMF) and vortexed for 30 seconds. Nafion® 117 membranes were cut into square strips approximately 0.5 inches a side. 50nm of gold was deposited onto the Nafion® strips by thermal evaporatoration (BOC Edwards Ebeam Evaporator). An adhesion layer or annealing were deemed unnecessary as gold adhered extremely well to Nafion®, even after hydroswelling of the proton exchange membrane. The fluoropolymer was then incubated with a 2-iminothiolane-labeled streptavidin compound for subsequent binding with a biotinylated BR. The gold-streptavidin layer functions as an orientation mechanism for biotin to attach. Purple Membrane was asymmetrically labeled on its extracellular side [3], with biotin, XX-SSE (6-((6-((biotinoyl)amino)hexanoyl)amino) hexanoic acid purchased from Sigma-Aldrich) for orientation and unidirectional pumping. Strips of Nafion® were affixed to a piece of silicon so that the gold-streptavidin side faced outwards from both sides of the silicon piece. The LB trough was cleaned as described above and the samples lowered into the subphase. Subsequently, DMF-solubilized PM was spread drop-wise at the air-water interface to a starting pressure of 12mN/m and allowed to equilibrate for one hour. Compression was performed in constant pressure mode to a deposition pressure of 30mN/m at a 100 Å^2/chain/min compression rate. The samples were then lifted at 1mm/min.

Adsorbance of asymmetrically-biotinylated PM fragments to streptavidin-coated Nafion® was confirmed by UV-vis spectrophotomety (Beckman). Proton conduction trials were performed by placing the samples in a two-chamber acrylic testing apparatus. The lower chamber was filled with a pH 4 solution, and the upper chamber with a pH 9 solution. The sample was placed with the extracellular sde of the purple membrane facing the pH 4 solution, forcing BR to pump against the flow of protons. Tests were run on samples without purple membrane, and on samples containing PM in light (250W Fostec lamp) and dark environments.

Gold-coated Nafion® strips were also interfaced with BR-polymer films. The LB trough was prepared similarly as above, Nafion® films lowered into the subphase, and polymer spread to a surface pressure of 15mN/m. Purified BR was injected to a pressure of 25mN/m. The barriers were compressed to 35mN/m and the sample lifted at 1mm/min. The samples were then tested for proton conductance in light and dark.

RESULTS AND DISCUSSION

By engineering our OmpF protein to include a cysteine residue at the Glu 183 site, we have a very powerful tool with which to label the protein for the purposes of orientation or visualization. OmpF is not a metalized protein, and thus cannot be seen explicitly under electron microscopy, but by conjugating the cysteine residue with 10nm colloidal gold beads we can obtain high-resolution images of protein distribution in a Langmuir film. Figure 1A shows a TEM image of a Langmuir film of triblock copolymer. After washing steps and blocking buffer baths to reduce non-specific binding, the sample was incubated with Nanogold® particles. Since the sample did not contain any protein, only one Nanogold® bead is seen, due to non-specific binding. Figure 1B shows a TEM image of a sample containing cys-tagged OmpF inserted into a Langmuir film of triblock copolymer. Dark spots represent the 10nm size

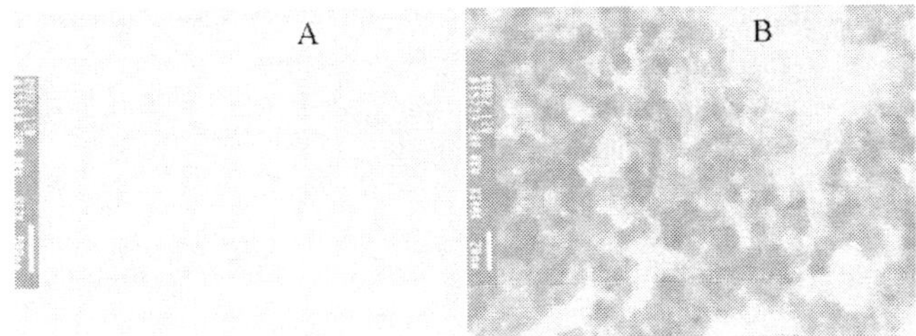

Figure 1. Polymer only shown in 1A. Cys-tagged OmpF conjugated with Nanogold particles shown in 1B.

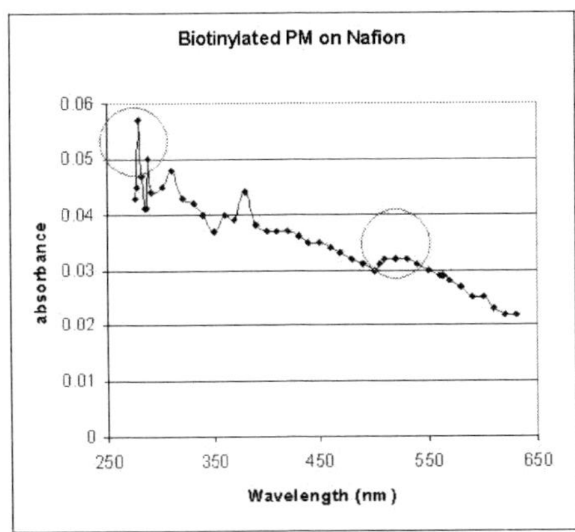

Figure 2. Characteristic absorbance peaks near 563 nm and 278nm of biotinylated PM adsorbed to Nafion.

gold particles, following wash steps, that have remained bound to the cys-tagged OmpF. During protein insertion, an immediate rise in surface pressure of typically 10mN/m is seen, and another 2-5mN/m increase is seen after two hours time has elapsed to allow for diffusion of remaining proteins to insert into the air-water interface. The presence of gold particles shown in Figure 1B confirms that protein is in fact making its way to the surface, and that the increase in surface pressure is not due to salts or detergent micelles.

UV-vis spectrophotometer readings, as seen in Figure 2, show characteristic retinal chromophore excitation peaks near 563nm and 278nm, indicating the presence of bacteriorhodopsin on Nafion®. Nafion®, as well as glass, quartz, silicon, and metals such as gold, is a suitable subtrate on which to deposit purple membrane fragments or polymer membranes. Where as Teflon® is extremely hydrophobic, the sulfonic acid side chains makes polymer extremely hydrophilic, thus allowing for monolayer films to be deposited on Nafion®. Detecting proton transport of BR in a Langmuir film requires access to protons from both sides of the film. Depositing such a monolayer of either PM or BR/polymer onto a solid substrate poses difficulty in testing for pH changes across the membrane. Porous substrates such as ultrafiltration membranes proved undesirable due to surface roughness that would impair complete film deposition. Nafion®, due to its versatility and conductance properties [4], addressed this problem. As a solid, robust proton transport material, Nafion® 117 eliminated the need for a porous material over which membrane formation would be extremely difficult. While its Teflon backbone is extremely hydrophobic, the sulfonic acid side chains make the flouropolymer hydrophilic and thus amenable to polymer or purple membrane deposition.

Another advantage of Nafion® is that gold can be easily evaporated onto it without an adhesion layer. Thicknesses between 10-50nm can easily be deposited without cracking or flaking, even under light manual abrasion and hydro-swelling of the Nafion® flouropolymer. Furthermore at 10-50nm in

Copyright © 2004 by ASME

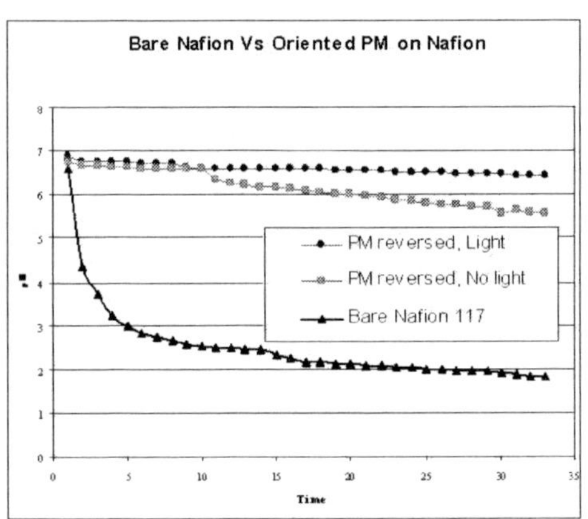

Figure 3. Proton Reversal trials across oriented hybrid protein/Nafion films in light and dark, compared to Nafion films without BR.

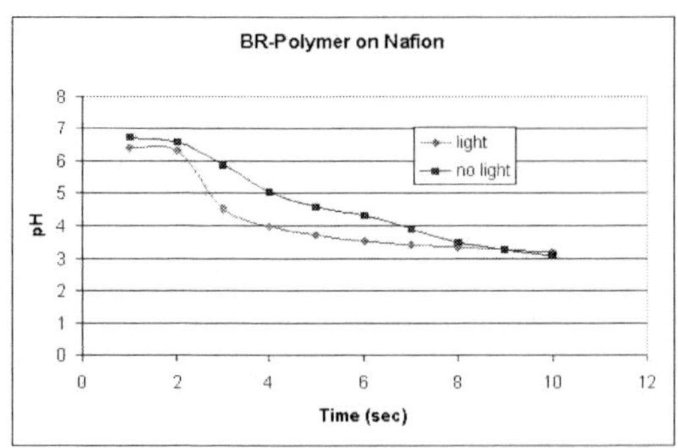

Figure 4. At the onset of the trial, a faster pH drop is seen with light than without, indicating faster proton transport across Nafion.

thickness, the gold is still porous enough to allow proton flow through the metal layer and light transmission.

The activity of DMF-solubilized biotinylated purple membrane, oriented onto a streptavidin/gold coated Nafion® film by Langmuir-Blodgett monolayer formation was tested by placing the film between a pH gradient. By asymmetrically labeling purple membrane with biotin in alkaline conditions, where the cytoplasmic lysine residues are not available to biotin, we can manipulate the membrane fragments so that only the extracellular side becomes biotinylated, consequently the extracellular side will interacted with streptavidin layer on the gold. One well of our testing chamber was filled with a pH 4 buffer, and the other well was filled with 100 μL pH 9 buffer. The oriented hybrid protein/Nafion® film was placed such that the extracellular side of the PM fragments faced the pH 4 buffer, forcing it to pump against the proton gradient. The pH probe was placed in the ph 9 buffer, but because of the sulfonic acid side chains of the proton exchange membrane, the effective pH is lower than 9 due to de-protonation of the Nafion® film. From figure 3, it can be seen that the initial pH was 6.59, 6.74, and 6.89 for trials ran for bare Nafion® film, the Nafion® interfaced with hybrid film, and the light-activated hybrid film, respectively. Due to the high acidity of the proton exchange membrane, scarcity of protons in 100μL of buffer of pH 9, the pH of the buffer is highly sensitive to even small increases in proton concentration. Without a hybrid film deposited on Nafion®, the pH of 100μL of buffer consistently drops below a value of 2 within two minutes of contact with Nafion®, but with a monolayer of oriented purple membrane/gold deposited on the flouropolymer, the donation of protons to the buffer is slowed. After 33 seconds, the bare film reached a pH of 1.82, the PM-enhanced film was at a pH of 5.55, and in the presence of light, the pH dropped to only 6.42. The difference in pH between darkness and exposure to light reached as high as 0.87 pH units

indicating function of light-dependent activity of BR protein on the hybrid film. The PM membrane provides a barrier to slow

the flow of protons from Nafion® to the buffer, and when exposed to light, proton flow down the gradient is reduced further, suggesting that we can not only reduce proton loss across proton exchange membranes, but to some extent, we can influence proton flow through light-actuated protein activity. The pH gradient between PM-enhanced Nafion® and Nafion® without protein was 3.73, and with the light-activated PM, the difference was as high as 4.6 after 33 seconds had elapsed. These gradients in 100μL of buffer correspond to the movement of 9.11×10^{17} protons from the Nafion® without protein, 1.59×10^{14} protons from the PM-enhanced Nafion® film, and 1.51×10^{13} protons upon light-activation. The amount of protons pumped by BR, which has a transport rate on the order of 100 protons per secion, is on the order of 10^9. BR monomers. Such high densities of protein can be expected due to nature's highly efficient expression of protein in purple membrane. The amount of energy generated in a fuel cell is dependent on the number of protons, as the reaction on the cathode is governed by

$$O_2 + 4H^+ + 4e^- \rightarrow 2H_2O.$$

The loss of protons will affect the efficiency of the reaction, and thus BR shows promise in facilitating proton transport, to prevent proton leakage, across proton exchange membranes for applicability in fuel cell technology. We have also observed that the 50nm of gold used for protein orientation was porous enough to facilitate the passage of protons across this film.

Proton leakage across the PM-enhanced film is likely due to unoriented fragments of PM that bound non-specifically to gold or portions of incomplete membrane coverage. The problem can be addressed by multiple Langmuir dips of the sample to effectively stack monolayer on top of one another. Multiple dips would lead to a more complete membrane coverage as well as increasing the density of active proteins in the z-axis.

The activity of a biomimetic system was measured in figure 4 In this case, BR was allowed to work with the pH gradient, as protons were expected to move from the high concentration buffer (pH 4) to the low concentration buffer (pH

9), but at different rates. Purified BR/polymer films were deposited onto Nafion® in a LB trough and then transferred to the testing chamber. Just as before, initial pH readings were lower than pH 9 due to the donation of protons from Nafion® into the buffer. In figure four, the pH value dropped faster with light than without light, indicating that in the presence of light, BR could be helping to transport protons from the PEM and into the buffer following the pH gradient. Eventually, both trials ended with pH values below 2, due to rapid flow of protons from the PEM and into the 100μL of buffer. It is believed that this mass exodus of proton transfer was not prevented due to incomplete film coverage.

CONCLUSION

ABA triblock copolymers represent a viable artificial membrane into which protein can be successfully inserted and reconstituted. By the engineering of OmpF to include a cysteine residue, we can confirm the insertion of protein into large-scale polymer membranes using a Langmuir-Blodgett trough. Protein insertion was confirmed by conjugation with colloidal gold and visualization under a TEM. We have also shown that unpurified bacteriorhodopsin in the form of purple membrane, oriented by asymmetric biotinylation, and purified BR inserted into block copolymer films can be successfully deposited on Nafion®, a proton exchange membrane. Upon light exposure, BR demonstrates active proton transport, which could improve the efficiency of the cathodic reaction in a fuel cell. Whereas protein density in the artificial system does not approach that of natural purple membrane, we are currently exploring methods to improve protein insertion at the air-water interface in LB monolayer formation, which may include increased diffusion time, varying salt/micelle concentrations, stirring of the subphase, or applying an electric field.

ACKNOWLEDGMENTS

The authors would like to thank Jinsoo Yi for the TEM imaging and colloidal gold preparation.

REFERENCES

[1] Hong, Q., Terrettaz, S., Ulrich, WP, Vogel, H. & Lakey, JH, "Assembly of pore proteins on gold electrodes," *Biochem. Soc. Trans.*, vol. 29, pp 578-82, 2001

[2] D. Ho, B. Chu, J. Schmidt, E. Brooks, C.D. Montemagno, "Hybrid Protein/Polymer Biomimetic Membranes" accepted to *IEEE Trans. Nanotechnology* June, 2004.

[3] R. Henderson, J.S. Stubb, S. Whytock, "Specific labelling of the protein and lipid on the extracellular surface of purple membrane," *J. Mol. Bio.*,vol. 123, pp. 259–274, 1978.

[4] K. Broka, P. Ekdunge, "Oxygen and hydrogen permeation properties and water uptake of Nafion117 membrane and recast film for PEM fuel cell,"*Journal of Applied Electrochemisty* vol.27 (2) 117-123, 1997.

NANO2004-46033

NANOMECHANICAL BIOSENSOR USING POLYMER MEMBRANES

Srinath Satyanarayana[1], Daniel T. McCormick[2] and Arun Majumdar[1,3]

[1]Department of Mechanical Engineering, [2]Berkeley Sensor and Actuator Center, Department of Electrical Engineering, University of California, Berkeley, CA 94720
[3]Materials Sciences Division, Lawrence Berkeley National Laboratory, Berkeley, CA 94720

In recent years several surface stress sensors based on microcantilevers have been developed for biosensing [1-4]. Since these sensors are made using standard microfabrication processes, they can be easily made in an array format, making them suitable for high-throughput multiplexed analysis. Specific reactions occurring on one surface (enabled by selective modification of the surface a priori) of the sensor element change the surface stress, which in turn causes the sensor to deflect. The magnitude and the rate of deflection are then used to study the reaction. The microcantilevers in these sensors are usually fabricated using material like silicon and its oxides or nitrides. The high elasticity modulus of these materials places limitations on the sensitivity and sensor geometry. Alternately polymers, which have a much lower elastic modulus when compared to silicon or its derivatives, offers greater design flexibility, i.e. allow the exploration of innovative sensor configurations that can have higher sensitivity and at the same time are suitable for integration with microfluidics and electrical detection systems.

A novel sensor array that uses polymer (parylene) micro-membranes has been developed to address the need for an accurate, rapid and quantitative tool for analyzing a variety of biomolecules like DNA, proteins and carbohydrates. The salient features of this sensor are: (i) it is label free (fluorescent/radio labels sometimes have a degrading effect on the reactions being studied); (ii) it is fabricated using a bio-compatible polymer; (iii) is designed for electrical and/or optical detection; and (iv) can be multiplexed easily.

The polymer micro-membrane sensor consists of a parallel plate capacitor in which one of the electrodes if fixed and the other electrode is a polymer membrane with a metal coating. Of the polymers that are currently used in microfabrication, parylene was chosen for sensor fabrication because it is biocompatible, produces a relatively stress-free film and is deposited at room temperature. Gold was chosen for the metal coating because it is compatible with the thiol chemistry used to attach the probe molecules to the sensor. The sensor is batch fabricated in an array format using standard microfabrication processes. A schematic of the sensor cross-section and a micrograph of the actual sensor are shown in Figure 1. The nitride layer below the bottom electrode isolates the different sensors in the microchip from each other. The microfluidic elements for sample I/O to individual sensors in the chip are fabricated on a glass chip which is bonded to the sensor chip using an ingenious adhesive bonding process that was developed in-house. The sensor and the corresponding microfluidics were specifically designed to prevent the liquid sample from entering the gap between the capacitor plates in order to avoid the undesirable effects of surface polarization and electric double layers. The glass chip also enables optical interrogation of the sensor element and thus offers an alternate method for verification.

The top electrode of the capacitor can deflect due to temperature changes like in a bi-material strip and change the sensor capacitance. Secondly, other factors like sample surface tension can also produce a false signal. These effects can be avoided by using a differential measurement setup, wherein, each fluid cell contains two sensor elements, one of which is a reference capacitor and the other is the sense capacitor (see Figure 1). The reference capacitor has its surface passivated to prevent reactions with the target molecules in the sample.

Coupled electro-mechanical analysis of the sensor element was performed using finite element method (ANSYS$^©$) to obtain the sensor response for a surface stress load and also for design optimization. For a given sensor geometry and surface stress load, the sensor response is a function of the relative size of the top electrode compared to the membrane size. The maximum sensor response was obtained when the top electrode size was 65% of the membrane size. The surface stress load used in the finite element simulation was 5 mJ/m^2 (the typical surface stress changes associated with bio reactions are in the order of 1-10 mJ/m^2 [1, 2]). The finite element model parameters and the corresponding sensor design parameters obtained from the simulation for an example are given in Table 1. The deflected shape of the sensor and the finite element mesh are shown in Figure 2.

The sensor chips are currently in the process of fabrication and the electronics circuit board for differential measurement is under development. Future work in this project involves characterization of the sensors and demonstration of multiplexed bio molecular interaction detection with DNA and proteins.

References:
[1] Yue, M., Lin, H., Dedrick, D. E., Satyanarayana, S., Majumdar, A. Bedekar, A. S., Jenkins, J. W., Sundaram, S., 2004, "A 2-D microcantilever array for multiplexed biomolecular analysis", Journal of Microelectromechanical Systems, **13(2)**, pp. 290-299.
[2] Wu, G., Datar, R., Hansen, K., Thundat, T., Cote, R. and Majumdar, A., 2001, "Bioassay of prostate specific antigen (PSA) using microcantilevers", Nature Biotechnology, **19**, pp. 856-860.
[3] Wu, G., Ji, H., Hansen, K., Thundat, T., Datar, R., Cote, R., Hagan, M. F., Chakraborty, A. K., Majumdar, A., 2001, "Origin of nanomechanical cantilever motion generated from biomolecular interactions", Proceedings of National Academy of Science, **98**, pp. 1560-1564.

[4] Fritz, J., Baller, M. K., Lang, H. P., Rothuizen, H., Vettiger, P., Meyer, E., Güntherodt, H. –J., Gerber, Ch., Gimzewski, G. K., 2000 "Translating biomolecular recognition into nanomechanics", Science, **288**, pp. 316-318.

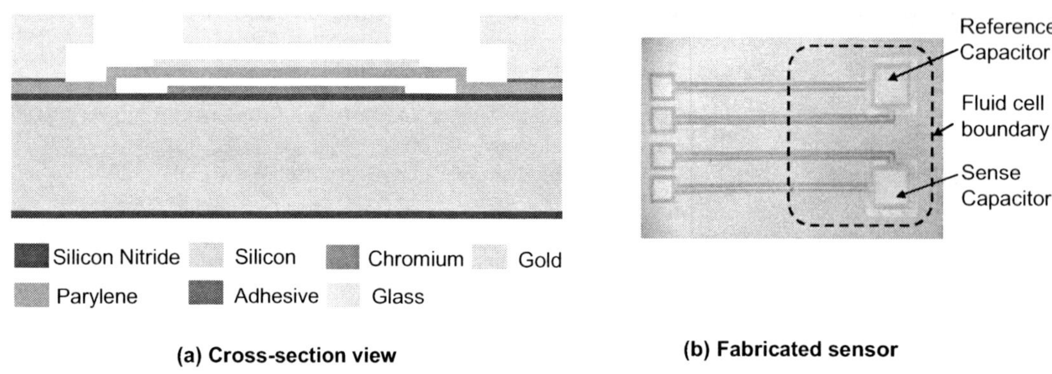

(a) Cross-section view **(b) Fabricated sensor**

Figure 1 Parylene membrane capacitor sensor

Table 1 Finite element simulation parameters and results

FE Simulation Parameters	
Membrane Size	300 μm square
Membrane thickness	0.5 μm
Gold thickness	0.025 μm
Top electrode size	240 μm
Bottom electrode size	220 μm
Surface stress change	5 mJ/m^2
Air gap	1 μm
FE Simulation Results	
Sensor capacitance	0.30 pF
$\Delta C/C$	2.12 %
Center deflection (δ)	34.4 nm
Thermal sensitivity	200 nm/K

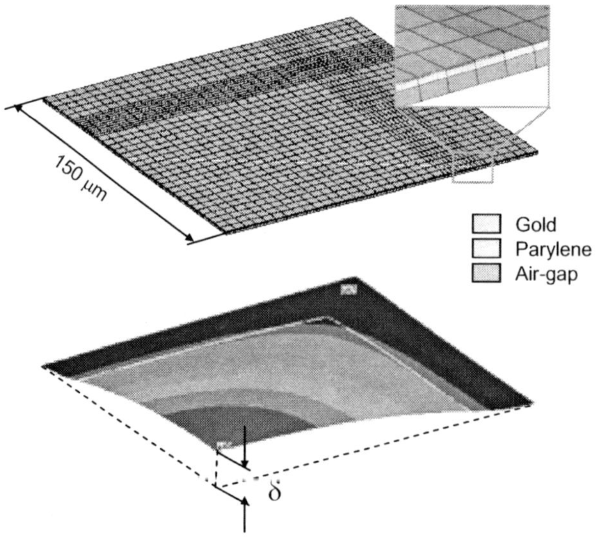

Figure 2 Finite element model (¼ symmetry) and the corresponding deflection contour plot for a surface stress change of 5 mJ/m^2.

NANO2004-46034

NANOMECHANICAL SENSOR ARRAY FOR DETECTION OF BIOMOLECULAR BINDINGS-TOWARD A LABEL-FREE CLINICAL ASSAY FOR SERUM TUMOR MARKERS

Min Yue
Department of Mechanical Engineering
University of California, Berkeley

Henry Lin
Department of Pathology
University of Southern California

Ram Datar
Department of Pathology
University of Southern California

Thomas Thundat
Oak Ridge National Laboratory

Richard J. Cote
Department of Pathology
University of Southern California

Jeanne C. Stachowiak
Department of Mechanical Engineering
University of California, Berkeley

Kenneth Castelino
Department of Mechanical Engineering
University of California, Berkeley

Karolyn Hansen
Oak Ridge National Laboratory

Arup Chakraborty
Department of Chemical Engineering
University of California, Berkeley

Arun Majumdar
Department of Mechanical Engineering
University of California, Berkeley

A label-free technique capable of rapidly screening human blood samples simultaneously for multiple serum tumor markers would enable accurate and cost-effective diagnosis of cancer before physiological symptoms appear. Recently, microfabricated, bimaterial cantilever sensors have been demonstrated to detect DNA hybridization and antigen-antibody binding at clinically relevant concentrations. Cantilever sensors deflect measurably under the surface stress resulting when biomolecules immobilized on one surface of the sensor interact with their binding partners [1]. We present an array of cantilever sensors (silicon nitride with a gold coated surface) capable of simultaneously interrogating 100 different biomolecular interactions.

Biomolecular interactions take place in individual microfluidic wells containing 4 to 12 cantilever sensors, the responses from which may be averaged (Figure 1A). The cantilever array chips are fabricated using a five-mask process based on conventional microfabrication techniques. A cap with microfluidic chambers, either etched into glass or molded into PDMS, separates neighboring wells within the array. Glass caps are bonded to the array using a UV-curing adhesive. PDMS caps

form a permanent bond with the array surface (silicon nitride) upon exposure to oxygen plasma. Whole-field laser illumination of rigid reflecting paddles on the cantilever sensor tips and collection of the resulting reflections by a CCD camera make simultaneous sensor readout possible. The cantilever paddles are imaged behind the focal plane so that paddle rotation results in image translation. The signal associated with a particular change in paddle angle increases as the camera is removed further from the image plane. However, the paddle images must be adequately focused to separate neighboring cantilever images, effectively limiting the defocus and determining the signal to noise ratio [1].

To demonstrate the 2-dimensional cantilever array chip as a tool for interrogating biomolecular interactions, we performed DNA Hybridization in a multiplexed format. Single-stranded, end-thiolated 20 base pair DNA was immobilized on the cantilever gold surfaces in 8 wells (about 32 cantilevers) simultaneously. The cantilevers were then monitored during exposure to an 8uM solution of non-complimentary 10 base pair DNA followed by exposure to a 5uM solution of fully complementary DNA. As Figure 1B shows, the cantilevers clearly deflected under DNA hybridization upon injection of the second sample. The sensor response was measured as the shift of each paddle image centroid in camera pixels. The shift of cantilever tips during a unit temperature change was recorded in the units of nanometers using a white light interferometer and compared to the image shift in pixels resulting from the same temperature change. Biological responses were normalized by the response to a unit temperature change and thus converted from pixels to nanometers. Finally a simple model for the relationship between tip deflection and surface stress in a bimaterial beam was used to convert tip deflection to surface stress change [1].

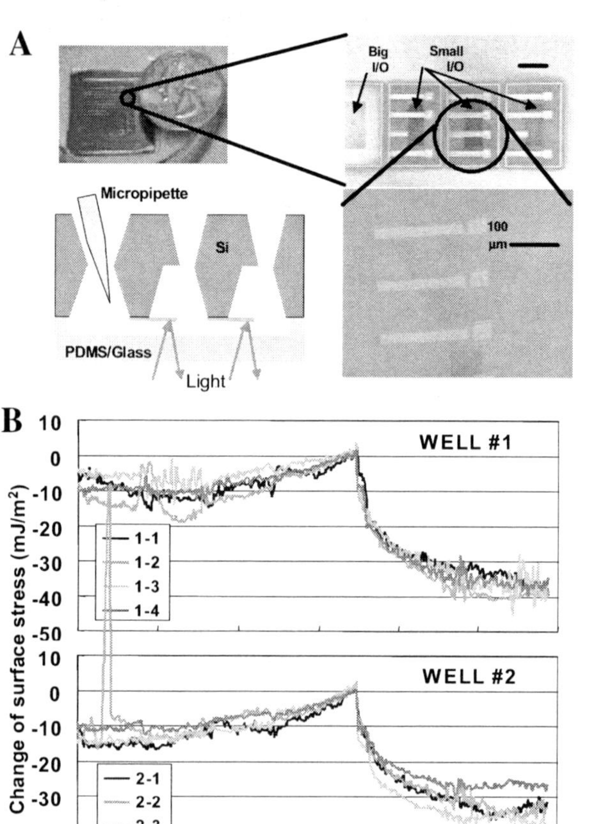

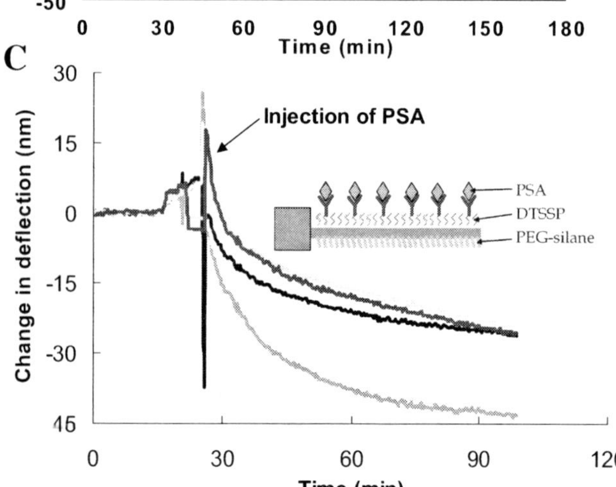

Figure 1. (A) Cantilever array chip, (B) DNA hybridization monitored by 2 groups of 4 cantilevers simultaneously, (C) Cantilever response to antigen-antibody binding

We study physical and chemical effects, such as solvent interactions, osmotic pressure, and strain energy responsible for cantilever deflection using the immobilization of thiolated DNA on gold-coated cantilevers and subsequent hybridization as a model system [2]. Experiments using fluorescent DNA (2µM) on gold surfaces were used to determine the immobilization and hybridization densities as a function of chain length (10, 20, 30 base pairs) and buffer concentration (10mM to 1M) The immobilization density increases with salt concentration and is controlled by osmotic and hydration forces in the low and high salt concentration regimes, respectively. However, high salt concentrations result in reduced hybridization efficiency due to steric effects. Taken together, these effects result in an optimal salt concentration where the hybridization density (product of immobilization density and hybridization efficiency) is maximized. Finally, the immobilization density also decreases for longer chain lengths. Current work focuses on the use of these results to control surface densities on cantilever sensors. As sensors of surface stess, cantilevers may provide a unique capability to examine the mechanical consequences of biochemical interactions.

We demonstrate that the cantilever arrays are capable of multiplexed specific antibody-antigen binding detection. Prostate specific antigen (PSA) and its specific antibody were chosen for the assay test, as PSA is an important serum tumor marker for prostate cancer. Polyethyleneglycol-silane (PEG-silane) was used to form a self-assembled layer on the silicon nitride side of the cantilever to block non-specific binding. The monoclonal antibody was immobilized through a cross-linker, 3,3'-Dithiobis(sulphosuccinimidyl propionate) (DTSSP), onto the gold surface of the cantilevers. The antibody oriented in random directions and served as the receptor to PSA. Figure 1C shows the change in the deflection of five cantilevers in two separate wells upon the injection of 20 µg/ml PSA, due to the specific binding of PSA to the antibody immobilized on the gold surface of the cantilevers. The variation in the change of different cantilevers' deflection may be due to the randomness of the probes on the surface and the variation of the local environment. PSA detection in lower concentrations and with the clinically relevant background is underway. We are currently attempting the use of PEG-thiol (less than 10 nm in length) to fill in open space on the gold surface as a potential means of preventing non-specific binding during exposure to the serum samples.

REFERENCES

[1] M. Yue, H. Lin, D. Dedrick, S. Satyanarayana, A. Majumdar, "A 2-D Microcantilever Array for Multiplexed Biomolecular Analysis," *Journal of Microelectromechanical Systems*, vol. 13, pp. 290-299, 2004.

[2] M. F. Hagan, A. Majumdar, and A. K. Chakraborty, "Nanomechanical forces generated by surface grafted DNA," *J. Phys. Chem. B* , vol. 106, pp. 10 163–10 173, 2002.

[3] G. Wu, R. Datar, K. Hansen, T. Thundat, R. Cote, and A. Majumdar, "Bioassay of prostate specific antigen (PSA) using

NANO2004-46037

ULTRA-HIGH FREQUENCY (UHF) NANOMECHANICAL RESONATOR INTEGRATED WITH PHASE LOCKED LOOP

X.L. Feng*, Y.T. Yang, C. Callegari, and M.L. Roukes
Electrical Engineering, Condensed Matter Physics, MC 114-36
California Institute of Technology, Pasadena, CA 91125

Nanoelectromechanical systems (NEMS) are interesting for both probing nanoscale physical fundamentals and exploring new technological applications [1]. In particular, nanomechanical resonators possess superb attributes including surprisingly-high operating frequency, ultra-small mass, high quality factor (Q), and thus are promising candidates for components in novel signal processing systems and ultra-sensitive sensors [1,2]. NEMS resonators with fundamental resonant frequencies exceeding 1GHz have been realized [3] and unprecedented mass sensitivity has also been demonstrated with VHF high-Q NEMS resonant mass sensors [2,4]. Among many engineering challenges to boost NEMS to more practical applications, it is of great importance to develop the generic protocol of integrating NEMS resonators with feedback and control systems. This work presents the first implementation of the integration of a UHF NEMS resonator with a low-noise phase locked loop (PLL).

The devices are fabricated from high quality SiC epilayer grown on Si substrate by APCVD, as SiC is proven to be wonderfully suitable for making RF and microwave NEMS resonators [3]. The AFM scan shows that the SiC layer surface roughness is about 1nm (Figure 1). As shown in Figure 2, the nanomechanical resonators are designed as doubly-clamped beams with in-plane fundamental flexural mode frequency in the UHF band. In the fabrication processes, the larger contact pads are patterned by photolithography technique, and the NEMS structure pattern is defined by e-beam lithography. The pattern is transferred and the resonator devices are suspended by a two-step ECR plasma etching with NF_3 and Ar, under carefully controlled composition, pressure and electrical biasing conditions. The devices have dimensions of 1.80um (length) × 150nm (width) × 100nm (thickness).

The NEMS resonators are tested with magnetomotive transduction. A balanced-bridge circuit detection technique is employed where a network analyzer (HP 8720C) drives a pair of impedance-matched resonator devices, sweeps the driving frequencies, and detects the resonance signals. Two UHF resonances at 417.2MHz and 419.9MHz, both with quality factor of $Q\sim1200$ have been detected for the two matched NEMS devices. Shown in Figure 3 is the electromechanical resonances signals seen by the first-stage amplifier in the front-end circuitry.

The 419.9MHz resonator is then integrated with a low-noise PLL which is specifically developed to interface with NEMS resonators, to perform real time frequency tracking and frequency stability characterization. The NEMS resonator's frequency shift with respect to the controlled changing of the device temperature is tracked in real time, as shown in Figure 4. The intrinsic frequency stability of the system, a crucial property of NEMS based oscillators and various NEMS based resonant sensors, is measured at stabilized temperatures. For instance, the measured Allan deviation (a measure of frequency instability [5]) is about 3×10^{-7} for 1.0 sec averaging time (Figure 5). The phase noise spectrum of the integrated NEMS and PLL system is also measured and analyzed (not shown in this abstract).

With the successful demonstration of the integration of NEMS resonator with PLL in the UHF range, exciting opportunities are anticipated for new advances in the development of NEMS based electrical oscillators, and various mechanical and biological sensors.

REFERENCES

[1] Roukes, M.L., "Nanoelectromechanical systems face the future", *Phys. World*, vol. **14**, pp. 25-31, 2001

[2] Ekinci, K.L., Yang Y.T., Roukes, M.L., "Ultimate limits to inertial mass sensing based upon nanoelectromechanical systems", *J. Appl. Phys.*, vol. **95**, pp. 2682-2689, 2004

[3] Huang, X.M.H., Zorman, C.A., Mehregany, M., and Roukes, M.L., "Nanodevice motion at microwave frequencies" , *Nature*, vol. **421**, pp. 496, 2003

[4] Ekinci, K.L., Huang, X.M.H., Roukes, M.L., "Ultrasensitive nanoelectromechanical mass detection", *Appl. Phys. Lett.*, vol. **84**, pp. 4469-4471, 2004

[5] Cleland, A.N., Roukes, M.L., "Noise processes in nanomechanical resoantors", *J. Appl. Phys.*, vol. **92**, pp. 2758-2769, 2002

---

* Presenting author. Email: xfeng@caltech.edu

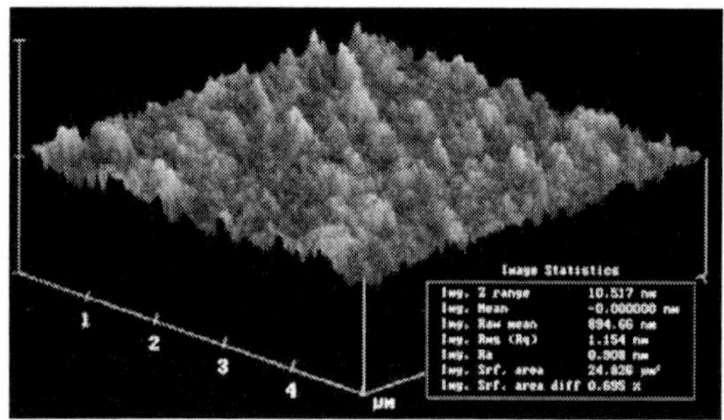

Figure 1. AFM image (5μm×5μm) of the SiC surface, with RMS surface roughness~1.15nm.

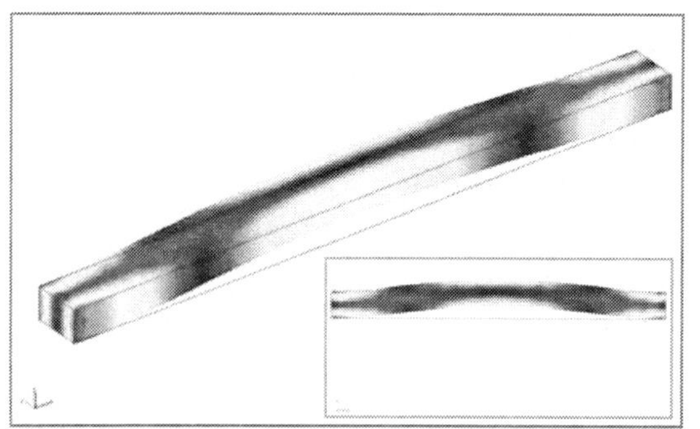

Figure 2. Mode shape and stress profile of the in-plane fundamental flexural mode of the doubly-clamped UHF NEMS resonator by finite element modeling. The inset is a top view of the in-plane mode shape in the x-y plane.

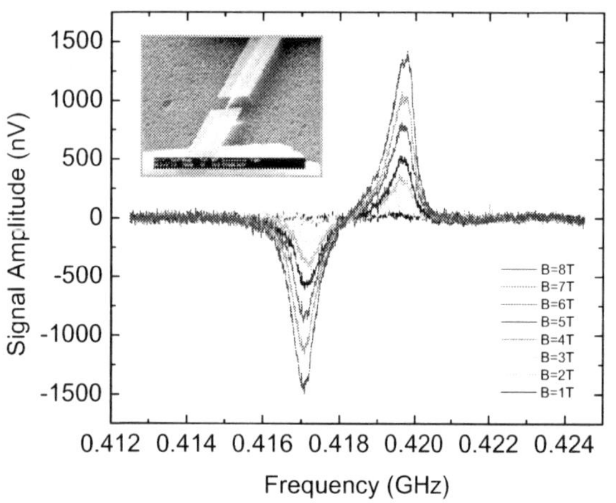

Figure 3. Measured electromechanical resonances of a pair of NEMS resonators, at 417.2MHz and 419.9MHz, respectively, both having Q~1200. The two devices are identically-fabricated, impedance closely-matched, and driven by out-of-phase (180° phase difference) RF signals split from a network

analyzer (HP 8720C). The plotted signal is referred to the input of the amplifier of the front-end circuitry. The balanced-bridge detection scheme is used in the measurement and the two branches of the bridge are carefully balanced. Inset is a SEM image of a UHF NEMS resonator, with measured dimensions: length $l{\approx}1.8$μm, width $w{\approx}150$nm, and thickness $t{\approx}100$nm.

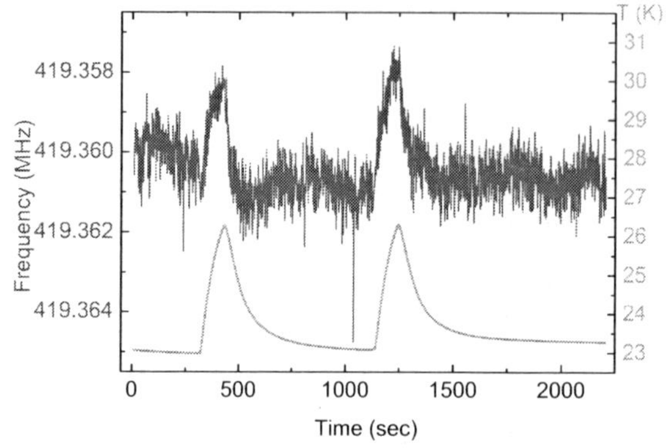

Figure 4. Demonstration of the temperature-programmed frequency shifting and frequency tracking with the phase locked loop (PLL): as the temperature is varied (red curve), the resonant frequency of the NEMS is shifted, and the frequency shifting is tracked (blue curve) by the PLL in real time.

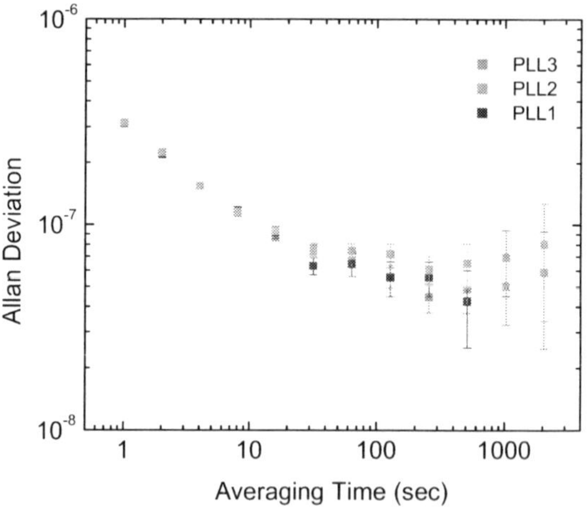

Figure 5. Measured frequency instability (Allan deviation) of the integrated NEMS-PLL system at stabilized temperature. Shown is the data for averaging time ranging from 1 sec to about 1 hour. The measured frequency instability characterizes the intrinsic phase noise level of the integrated NEMS resonator and PLL system.

NANO2004-46047

Nanoscale Heat Conduction in Data Storage Technology

Mehdi Asheghi

Mechanical, Engineering Department, Carnegie Mellon University, Pittsburgh, PA 15213

The magnetic data storage industry has followed a similar density (and data rate) improvement curve as the semiconductor technology (Moore's Law) for the past decade. However, whether the storage densities will continue to increase at this rate and be able to keep up with the improvements in processor technology is under a near term threat resulting from the fundamental physics up on which the hard disk drives are based. It is expected that novel, more unconventional technological solutions become necessary to overcome limitations, however, many of these technologies rely heavily on heating and energy transport at extremely short time and length scales. It is widely believed that further advances in high-technology data storage systems will be difficult, if not impossible, without rigorous treatment of the nano-scale energy transport. The nano-scale heat transfer research effort at Data Storage System Center (DSSC) has been focused on three interwoven areas of thermal design, failure analysis, and metrology of micro/nano-devices and structures relevant to data storage technologies. In this presentation, underlying physics and fundamentals of heat transport at nanoscale will be discussed. In addition, applications of the nanoscale heat transfer to the thermal analyses of the magnetic and phase change optical data storage technologies will be presented.

The problem facing the HDD industry is due to the continued shrinkage of the dimensions of the bit cells on the disk. Each of these individual magnetic data bit cells is comprised of a collection of smaller crystalline grains of a thin magnetic film, each of which is uniformly magnetized. In order to maintain adequate signal-to-noise ratio for reliable data recording and retrieval, it is necessary to keep the number of such grains per bit cell adequately large. This has required the size of the grains to be reduced with increasing AD. The superparamagnetic effect becomes important when the grain volume V is so small, that the inequality $K_u V/k_B T > 40$ can no longer be met [1]. Here K_u is the material's magnetic crystalline anisotropy energy density, k_B is the Boltzmann's constant, and T is absolute temperature. When the above inequality is not satisfied, thermal energy demagnetizes the individual grains and the stored data bits will not be stable. Therefore, as we make the grain diameter smaller with increasing AD, we reach a regime for a given material K_u and T, such that reliable data storage is no longer feasible. With the current pace of miniaturization, some experts believe the industry could reach this barrier as early as 2005. Data storage opportunities using nanoscale thermal effects arise within system architectures that overcome the superparamagnetic effect by using thermally assisted writing schemes (e.g., hybrid thermomagnetic disk recording systems, thermally assisted scanned probe magnetic recording systems, and cantilever probe based thermomechanical recording systems) [2]. The data storage industry's ambition to improve AD and data rates has pushed the characteristic length and timescales of these devices and processes down to the nanometer and picoseconds range, respectively. If a small quantity of heat is generated over a period of time less than the thermal time constant of the magnetic element or device, then heat travels only a few nanometers away during the heating pulse. The energy must therefore be absorbed by a very small volume, which dramatically increases the temperature in the device, structure, or system. Such heating phenomena will be vastly underestimated by continuum theory and could melt the magnetic element, otherwise damage it, or cause long-term reliability concerns.

The fundamentals of energy transport (by phonon, electron, lubricant molecules and gas molecules) in nanostructures and nanoscale components of the data storage devices/systems are not well understood at present. Further advances in the relevant high-density data storage systems will be difficult, if not impossible, without rigorous treatment of energy transport at extremely short time and length scales. This manuscript shows that the fundamental knowledge developed through nanoscale thermal characterization and research can be used to realize the technological opportunities and addresses the technological threats in data storage devices and systems.

References
[1] Charap, S. H., Lu, P. L., and He, Y., 1997, "Thermal Stability of Recorded Information at High Densities," *IEEE Tran. Mag.,* Vol. 33, pp 978-983.
[2] Asheghi, M., Yang, Y., Sadeghipour, M.S., Bain, J.A., Barmak, K., Jhon M.S., Schlessinger, T.E., Zhu, J.G., and White, R.M., 2002, "Nanoscale Energy Transport in Information Technology Research with an application to High-density Data Storage Devices and Systems," ASME International Mechanical Engineering Congress & Exposition, Invited Paper No. 2-16-1-14, November 17-22, 2002, New Orleans, Louisiana.

NANO2004-46050

Nanoscale Heat Conduction in the SOI, Strained-Si and Tri-Gate Transistors

Mehdi Asheghi

Mechanical, Engineering Department, Carnegie Mellon University, Pittsburgh, PA 15213

There have been many attempts in the recent years to improve the device performance by enhancing carrier mobility by using the strained-induced changes in silicon electronic bands [1-4] or reducing the junction capacitance in silicon-on-insulator (SOI) technology. Strained silicon on insulator (SSOI) is another promising technology, which is expected to show even higher performance, in terms of speed and power consumption, comparing to the regular strained-Si transistors. In this technology, the strained silicon is incorporated in the silicon on insulator (SOI) technology such that the strained-Si introduces high mobility for electrons and holes and the insulator layer (usually SiO_2) exhibits low junction capacitance due to its small dielectric constant [5, 6]. In these devices a layer of SiGe may exist between the strined-Si layer and insulator (strained Si-on-SiGe-on-insulator, SGOI) [6] or the strained-Si layer can be directly on top of the insulator [7]. Latter is advantageous for eliminating some of the key problems associated with the fabrication of SGOI.

However, self-heating and electrostatic discharge (ESD) events have become serious performance and reliability concerns for SOI and strained-Si transistors, which are separated from the silicon substrate by poor thermal conducting layers [8-11]. In addition, it is well established that thermal conductivities of silicon layers of thickness less than 300 nm are significantly less than the bulk values, due to the phonon-boundary scattering [12,13]. This can further impede heat conduction from the device and subsequently exacerbate the self-heating problem.

While the effect of phonon-boundary scattering have been investigated in the past for thin silicon layers in the order of ~ 1 μm at room and cryogenic temperatures [13,14], very limited experimental data are available for silicon layers of thickness 10-100 nm and at room temperature and above. We have recently succeeded to measure thermal conductivity of the silicon layers of thickness 20 and 100 nm at high temperatures [15]. These data along with the predictions for phonon transport using thermal conductivity integral in relaxation time approximation and proper modification for the reduced mean free paths, due to phonon boundary scattering [13,16], are given in Fig. 2. In addition, the graph depicts the thermal conductivity data for individual silicon nanowires of various diameters [16], which agree well with the predictions of phonon-boundary scattering.

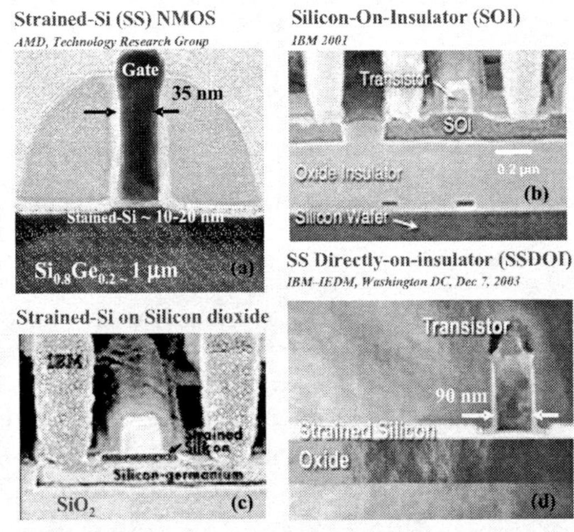

Figure 1: Scanning Electron Micrograph (SEM) of the (a) Strained-Si, (b) silicon-on-insulator, (c) strained-Si on-insulator and (d) strained-Si on insulator (SSOI) transistors. The active region of these devices are made of thin silicon layers of thickness in the order of 10-100 nm, which are separated from the silicon substrate by poor thermal conducting layers.

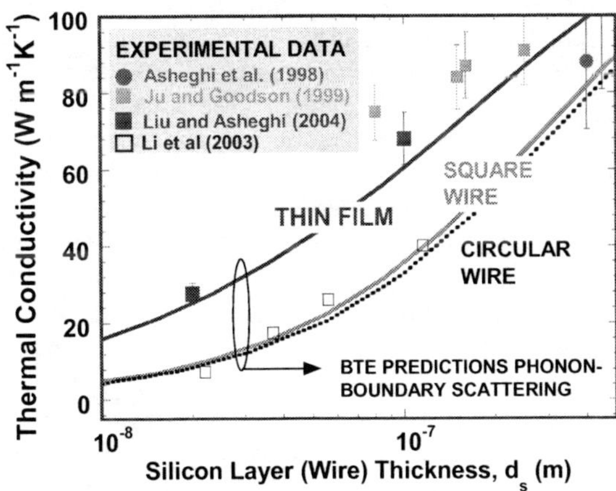

Figure 2: Thermal conductivity of silicon layer and wires, at room temperature, as a function of thickness. Predictions based on Boltzmann transport equation (BTE) agree reasonably well with the experimental data [13, 15, 16].

It is customary to define a thermal resistance for electronic devices in order to determine an average or maximum temperature rise at the device for a given average dissipated power [17]. According to the roadmap of semiconductor industry, ITRS, the gate length of transistors in the year 2004 is as small as 60 nm. Recalling that the thickness of the silicon in an SOI device should be 1/3 of the gate length to maintain full depletion [18], one can conclude that the thickness of Si layer has to be around 20 nm. Using the Lumped Analytical (LA) model [17], the dc thermal resistance of SOI transistors as a function of the device thickness is predicted and shown in Fig. 3. It is clear that neglecting the effect of phonon-boundary scattering by using the bulk thermal conductivity value for silicon can result in a significant error in calculation of the SOI device thermal resistance.

The LA (multi-fin) model [17] is adapted here to predict the thermal resistance of tri-gate transistors. The thermal conductivity of square nano-wires (Fig. 2) is used to describe the thermal transport in the active region of tri-gate transistors. As shown in Fig. 4, thermal resistance of tri-gate transistor is a strong function of silicon film thickness. An interesting comparison can be made between the thermal resistance of SOI devices and that of tri-gate transistors. Let us compare the self-heating of a 60nm gate length SOI device with a 60nm gate length tri-gate transistor. In a typical SOI transistor with 60nm gate length, the acceptable thickness for silicon layer for fully depleted transistor is ~ 20 nm [18] whereas in a tri-gate transistor device layer thickness of ~ 60nm is sufficient to achieve full depletion condition. We shall examine the typical values of the thermal resistance **per micron of device width** as well as drain current per unit width for a given voltage in order to make a comparison between these two technologies. The latter determines the amount of heat generation per unit width of the device. Using the measured data and an analytical expression for I-V characteristic of SOI devices [19,20] the saturation current of an SOI transistor with gate length, L_g, of 60 nm is estimated to be 0.4-1 mA/µm at

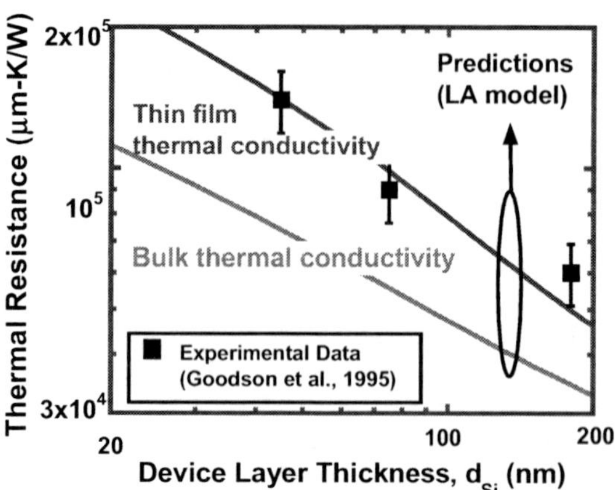

Figure 3: Thermal resistance of SOI transistor as a function of the silicon layer thickness. Experimental data obtained from a 10µm wide transistor with 360 nm oxide thickness and 100nm gate length (Goodson et al., 1995).

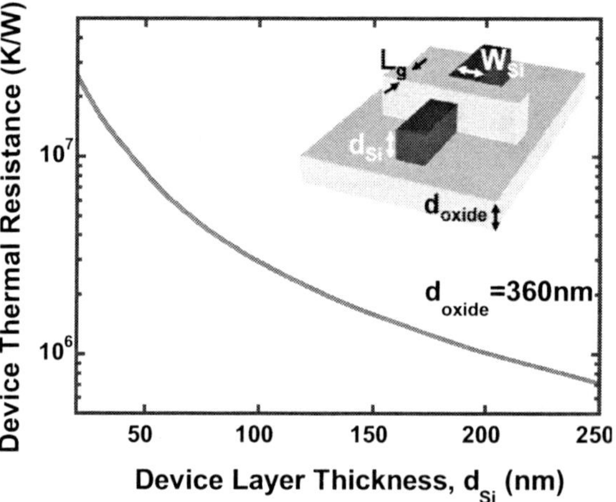

Figure 4: Thermal resistance of tri-gate transistors as a function of Si thickness. Width (W_{Si}), gate length (L_g), and thickness (d_{Si}) of the device were assumed to be equal (Doyle, 2003). Oxide thickness is 360nm.

drain voltage equal to 1V. The saturation current of 0.52 mA/µm is reported for a tri-gate transistor with L_g=60nm at drain voltage of 1V [17]. Having the gate length of 60nm, the thermal resistance of the SOI device is equal to 2.3×10^5 µm-K/W (device thickness=20nm, Fig. 3). Choosing d_{Si} and L_g to be 60nm, the thermal resistance of the tri-gate transistor is estimated to be 3.6×10^5 µm-K/W. Therefore, comparing to SOI devices, tri-gate transistors experience higher level of self-heating. For both cases, it was confirmed that the distance between interconnects and gate is considerably larger than the healing length in the devices and as the result the effect of interconnects on the overall thermal resistance is neglected. A complete comparison of self-heating should include electro-thermal simulations with special attention to the impact of non-uniform heat generation in the channel region, heat spreading in the device, temperature dependency of mobility and threshold voltage.

The steady-state modeling of self-heating in SOI devices can be readily extended to the strained-Si transistors. This model takes advantage of the fact that the thick $Si_{0.8}Ge_{0.2}$ underlayer effectively separates the device from the silicon substrate such that 98% of the temperature rise occurs within the starined-Si and SiGe underlayer. In addition, the contrast in thermal resistances along the strained-Si layer, R_{si}, and SiGe underlayer, R_{SiGe}, allow us to decouple and simplify the heat conduction in the device to the multi-fin one-dimensional heat conduction configuration [17]. Figure 5 compares the self-heating experimental data [20] with predictions of the multi-fin model and finite element simulations [10]. The uncertainty for data is estimated to be near ~20% but it is very difficult to precisely assess the errors without the exact knowledge of the calibration process. The slope of the curve that passes through most of the experimental data yields the device thermal resistance of 1.63×10^4 K/W and agrees well with ANSYS finite element simulations. The predictions

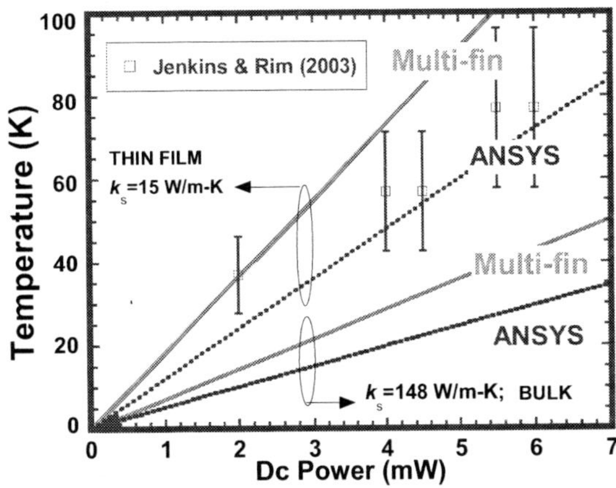

Figure 5: Experimental data [20] and predictions of self-heating in a strained-Si transistor [10].

based on the multi-fin model, 1.89×10^4 K/W, is about 15% higher than the ANSYS simulations. One should have in mind that the multi-fin model is best suited for parametric study and scaling analysis of the heat transport in strained-Si devices and that no fitting parameter has been used in these predictions. If necessary, the model can be further calibrated against the experimental data but considering the uncertainty in the experiments this approach may not be warranted. The impact of phonon-boundary scattering or reduce thermal conductivity of silicon is depicted in the same graph. The results indicate that the estimated thermal resistances are 5×10^3 K/W and 8×10^3 K/W for ANSYS and multi-fin model, respectively. These values are nearly a factor of two smaller than those predicted by the sub-continuum model presented in this work.

References

[1] J. Welser, J.L. Hoyt, J.F. Gibbons, "NMOS and PMOS Transistors Fabricated in Strained Silicon/Relaxed Silicon-Germanium Structures," IEEE IEDM, pp.1000-2, Dec. 1992.

[2] K. Rim, S. Koester, M. Hargrove, J. Chu, P.M. Mooney, J. Ott, T. Kanarsky, P. Ronsheim, M. Ieong, A. Grill, H.-S.P Wong, "Strained Si NMOSFETs for High Performance CMOS Technology," VLSI Technology Digest of Technical Papers Symposium, pp. 59-60, June 2001.

[3] M. Rashed, S. Jailepalli, R. Zaman, W. Shih, T.J.T. Kwan, C.M. Maziar, "Simulation of Electron Transport in Strained Silicon on Relaxed $Si_{1-x}Ge_x$ Substrates," Proc. Eleventh Biennial University/Government/Industry Microelectronics Symposium, pp. 168-171, May 1995.

[4] H. Miyata, T. Yamada, D.K. Ferry, "Electron Transport Properties of a Strained Si Layers on a Relaxed $_{-x}Ge_x$ Substrates by Monte Carlo," Applied Physics Letters, vol. 62(21), pp. 2661-2663, 1993.

[5] L. Haung, J.O. Chu, S.A. Goma, C.P. D'Emic, S.J. Koester, D.F. Canaperi, P.M. Mooney, S.A Cordes, J.L. Speidell, R.M. Anderson, H.-S.P. Wong, "Electron and Hole Mobility Enhancement in Strained SOI by Wafer Bonding," IEEE Transaction on Electron Devices, vol. 49(9), pp. 1566-1571, Sept. 2002.

[6] T. Tezuka, N. Sugiyama, T. Mizuno, S. Takagi, "High-Performance Strained Si-on-Insulator MOSFETs by Novel Fabrication Processes Utilizing Ge-Condensation Technique," VLSI Technology Digest of Technical Papers Symposium, pp. 96-97, June 2002.

[7] T.A. Langdo, A. Lochtefeld, M.T. Currie, R. Hammond, V.K. Yang, J.A. Carlin, C.J. Vineis, G. Braithwaite, H. Badawi, M.T. Bulsara, E.A. Fitzgerald, "Preparation of Novel SiGe-Free Strained Si on Insulator Substrates," IEEE International SOI Conference, pp. 221-222, Oct. 2002.

[8] K. Rim, J. Chu, H. Chen, K.A. Jenkins, T. Kanarsky, K. Lee, A. Mocuta, H. Zhu, R. Roy, J. Newbury, J. Ott, K. Petrarca, P. Mooney, D. Lacey, S. Koester, K. Chan, D. Boyd, M. Ieong, H.-S. Wong, "Characteristics and Device Design of Sub-100 nm Strained Si N- and PMOSFETs," VLSI Technology Digest of Technical Papers Symposium, pp. 98-99, June 2002.

[9] J.L. Hoyt, H.M. Nayfeh, S. Eguchi, I. Aberg, G. Xia, G., T. Drake, E.A. Fitzgerald, D.A. Antoniadis, "Strained Silicon MOSFET Technology," IEEE IEDM, pp. 23-26, Dec. 2002.

[10] W. Liu, M. Asheghi, "Thermal Modeling of Self-Heating in Strained-Silicon MOSFETS", submitted to ITherm Conference, NV, June 2004.

[11] K. Etessam-Yazdani, M. Asheghi, "Ballistic Phonon Transport on Strained Si/SiGe Nanostructures with an Application to Strained-Silicon Transistors," submitted to ITherm Conference, NV, June 2004.

[12] Y.S. Ju, "Microscale Heat Conduction in Integrated Circuits and their Constituent Films," Ph.D. Thesis, Stanford University, 1999.

[13] Asheghi, M., Touzelbaev, M.N., Goodson, K.E., Leung, Y.K., and Wong, S.S., "Temperature Dependent Thermal Conductivity of Single-Crystal Silicon Layers in SOI Substrates," J. of Heat Transfer, vol. 120, pp. 30-36, 1998.

[14] M. Asheghi, K. Kurabayashi, R. Kasnavi, and K. E. Goodson, "Thermal Conduction in Doped Single-Crystal Silicon Films," Journal of Applied Physics, vol. 91(8), pp. 5079-88, Apr. 2002.

[15] W. Liu, S. Zhang, Y. Yang, S. Sadeghipour, M. Asheghi, "Thermal Conductivity of the 100 nm Single Crystal Silicon Film," Proc. ASME International Mechanical Engineering Conference and Exposition, Paper No. IMECE2003-41628, Nov. 2003.

[16] D. Li, Y. Wu, P. Kim, L. Shi, P. Yang and Arun Majumdar, 2003, "Thermal conductivity of individual silicon nanowires," Journal of Applied Physics Letters, vol. 83 (14), pp. 2934-36.

[17] Goodson, K.E., Flik, M.I., Su, L.T., and Antoniadis, D.A., 1995, "Prediction and Measurement of Temperature Fields in Silicon-on-Insulator Electronic Circuits," *ASME Journal of Heat Transfer*, vol. 117, pp. 574-581.

[18] Doyle, B.S., Datta, S., Doczy, M., Hareland, S., Jin, B., Kavalieros, J., Linton, T., Murthy, A., Rios, R., Chau, R., 2003, "High performance fully-depleted tri-gate CMOS transistors," *IEEE Electron Device Letters*, vol. 24(4), pp. 263-265

[19] Hsiao, T.C., Kistler, N.A., Woo, J.C.S., 1994, "Modeling the I-V characteristics of fully depleted submicrometer SOI MOSFET's," *IEEE Electron Device Letters*, vol. 15(2), pp. 45-47

[20] Jenkins, K.A., Franch, R.L., 2003, "Impact of self-heating on digital SOI and strained-silicon CMOS circuits, " *IEEE International SOI Conference*, pp.161-163

NANO2004-46051

MULTIPLEXED LABEL-FREE BIOSENSOR CHIP BASED ON NANOSCALE ELECTRICAL DOUBLE LAYER CAPACITANCE SENSING

Kenneth Castelino, UC Berkeley, Mech. Engg.

Veljko Milanovic, UC Berkeley, Mech. Engg.

Daniel T. Mc.Cormick, UC Berkeley, Electrical Engg.

Norman Tien, UC Davis, Electrical Engg.

Arun Majumdar, UC Berkeley, Mech. Engg.

ABSTRACT

We report the design, fabrication and testing of a microchip that exploits a capacitive detection scheme for multiplexed label-free biomolecular assays. The detection scheme is based on the nanoscale gap parallel-plate capacitor formed by the electrical double layer at the interface of a metal electrode and an ionic solution, which is sensitive to biological reactions at the electrode surface. Since the nanogap is obtained by electrical and chemical control, no nano-patterning techniques are needed and the simple device structure facilitates sensor readout, multiplexing, and packaging. Finally, this sensing technique is universal and can be applied to detection of diverse biological entities such as proteins and cells.

INTRODUCTION

Currently, most biomolecule detection schemes involve use of fluorescent, magnetic, chemiluminescent, or other labels. While these detection techniques are very sensitive and also quantitative, they require extra sample preparation or assay steps that are often difficult or undesirable. Label-free detection techniques monitor changes in electronic, optical, magnetic, or other properties that occur during a bimolecular reaction without the need for labeling the sample being analyzed. A large class of label-free sensors based on surface-plasmon resonance [1], optical waveguides [2], acoustic resonators [3], and electrochemical phenomena [4] are sensitive to reactions at surfaces and have been demonstrated in a number of different assays involving DNA, proteins, peptides and cells. The main challenge for such sensors is detection in complex mixtures such as serum, where non-specific binding is a major issue.

The electrochemical properties of metallic and semiconductor electrodes in contact with ionic solutions have been extensively characterized and various techniques based on sensing changes in electrode capacitance or Faradaic currents due to biological recognition events have been demonstrated [5]-[6]. However, these macroscale setups are difficult to multiplex, require large sample volumes and are not suited for applications such as cancer diagnostics, which require detection of multiple targets. We have developed a capacitive biosensor that senses changes in the electrical double layer at the electrode surface due to biomolecular reactions. The device has multiple sensing elements that are multiplexed electronically and can be used to detect multiple targets simultaneously.

THEORY

An electrical double layer is formed at the interface of a metal electrode and an ionic solution due to accumulation of ionic species at the interface. This double layer acts as a parallel-plate capacitor with a characteristic gap on the order of a few nanometers, which can be tuned by changing the applied bias voltage and the concentration of the ionic species. The double layer capacitance is highly sensitive to reactions such as immobilization or binding of biomolecules that occur at the electrode surface. The capacitance is measured using an impedance technique, where a small sinusoidal voltage is applied in addition to the bias voltage and the impedance of the system is measured. The measured impedance can be modeled using the Randles equivalent circuit [7] that accounts for the faradaic resistance, the diffusion impedance, the solution resistance, and the double layer capacitance. In the frequency range of interest (200 Hz – 10 kHz), the dominant factors are the solution resistance and double-layer capacitance and the equivalent circuit reduces to a simple series R-C circuit.

EXPERIMENTS

An optical micrograph and a schematic of the microchip are shown in Fig. 1a and Fig. 1b. Each sensor element consists of a gold sensing and counter electrode. The counter electrode area is much larger so that the dominant capacitance in the system is due to the sensing electrode. The microchip integrates a fluidic sample chamber, a microheater for fast temperature cycling up to 100°C. The entire chip is wire bonded into a package, which is connected to a multiplexing circuit that can access any of the devices. The system impedance is measured using a HP4194A impedance analyzer and a series R-C circuit model is used to extract the capacitance.

The nanogap thickness and device sensitivity depend on the ionic concentration, which was investigated by monitoring capacitance changes during monolayer formation of the alkanethiol mercaptohexanol (MCH) at two different buffer concentrations (20 / 200 mM) as shown in Fig 1(c). The capacitance decreases in both cases due to the displacement of water and ions from the self-assembled monolayer (SAM). However, the percentage capacitance change for the 20 mM case (58%) was less than that for the 200 mM (83%) since the double-layer is thicker and hence is less sensitive to the SAM. After the immobilization reaction, the chips were cleaned in O_2

plasma in order to remove the immobilized molecules. After the plasma treatment, the capacitance returned to near the original value. The discrepancies noted could be due to changes in the gold surface roughness during the cleaning process.

The capacitance changes due to immobilization of a 20 base-pair single-stranded DNA (ss-DNA) and hybridization with the complementary DNA is shown in Fig 2a. Both reactions reduce capacitance due to the displacement of water from the double layer region by the biomolecules. In order to reduce the non-specific binding, the electrodes were passivated using a layer of MCH after immobilizing the thiolated probe DNA. The MCH insulates the electrode and removes non-specifically adsorbed DNA from the surface. We also used a reference device passivated with MCH but not functionalized with the probe DNA to monitor the amount of non-specific binding and demonstrated detection of DNA hybridization at 50 nM concentration in a 10 microliter sample volume as shown in Fig 2(b). Finally, the integrated microheater was used to study the DNA melting and rehybridization process for DNA as seen in Fig 2(c). Heating the hybridized DNA beyond the melting point leads to de-hybridization causing capacitance to increase to the immobilization value. After 15-20 minutes, the re-hybridization process causes capacitance to return to the original hybridization value.

DISCUSSIONS AND CONCLUSIONS

We have demonstrated a multiplexed capacitive biosensor that can be used for probing biomolecular reactions involving DNA, proteins, and cells. We are currently investigating electrode passivation schemes based on self-assembled monolayers in order to minimize non-specific binding and improving the sensor resolution for single base-pair mismatches detection. Finally, this technique is currently being extended to antibody-antigen detection for detecting multiple targets in complex environments such as serum.

REFERENCES

[1] K.A. Peterlinz, R.M. Georgiadis, T.M. Herne, and M.J. Tarlov, *J. Am. Chem.Soc.*, **119**, pp. 3401, 1997.

[2] M. Pawlak, *et al*, *Proteomics*, **2**, pp. 383-393, 2002.

[3] A.W. Wang, R. Kiwan, R.M. White, and R.L. Ceriani, *Sensors & Actuators B-Chemical*, **49**, pp. 13-21, 1998.

[4] E. Katz, and I. Willner, *Electroanalysis*, **15**(11), pp. 913-947, 2003.

[5] C. Berggren, P. Stalhandske, J. Brundell and G. Johansson, *Electroanalysis*, **11**(3), pp. 156-160,1999.

[6] F. Patolsky, *et.al.*, *J. Phys. Chem B*, **102**, pp. 10359, 1998.

[7] A. Bard, and L. Falkuner, *Electrochemical Methods'*, John Wiley & Sons, Inc. (2001).

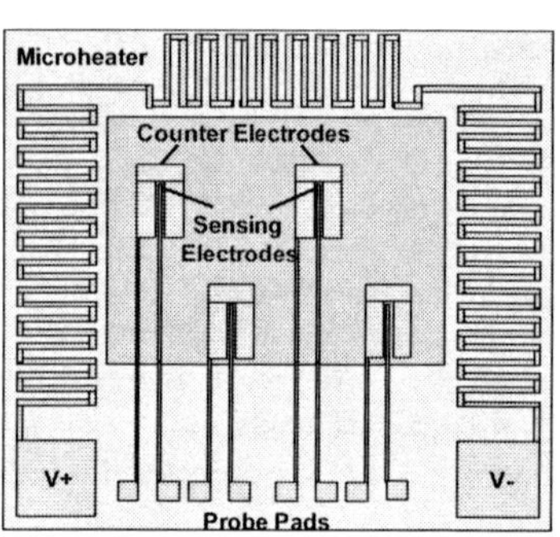

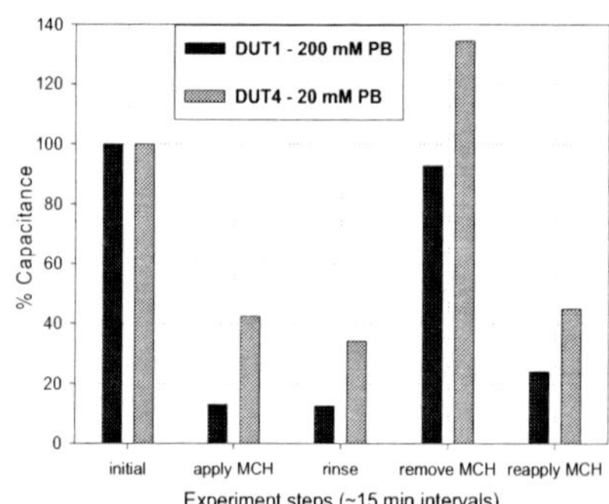

Fig. 1. Microchip for multiplexed capacitive biosensing: (a) optical micrograph of the chip wire bonded to the external package; (b) Schematic showing the layout of sensing elements; (c) Capacitance changes (normalized) due to immobilization of mercapto hexanol on gold electrodes. After cleaning with O₂ plasma, the sensing elements were restored to their original condition and reused.

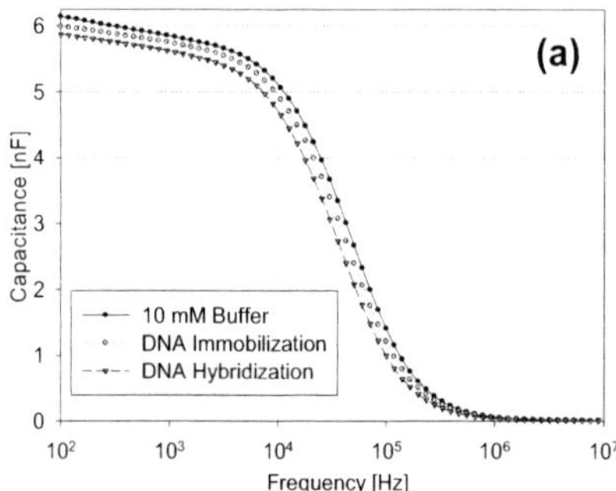

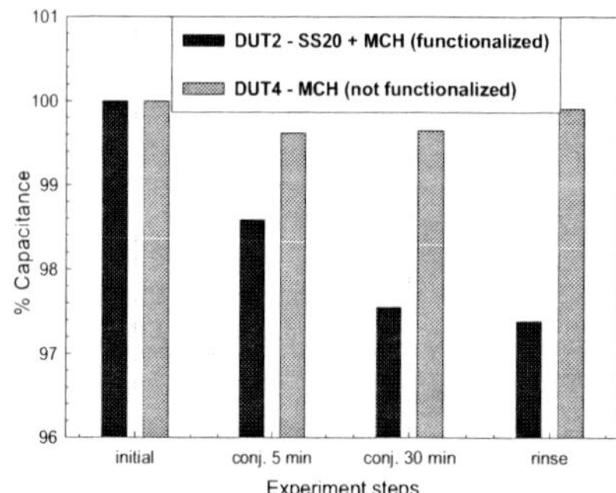

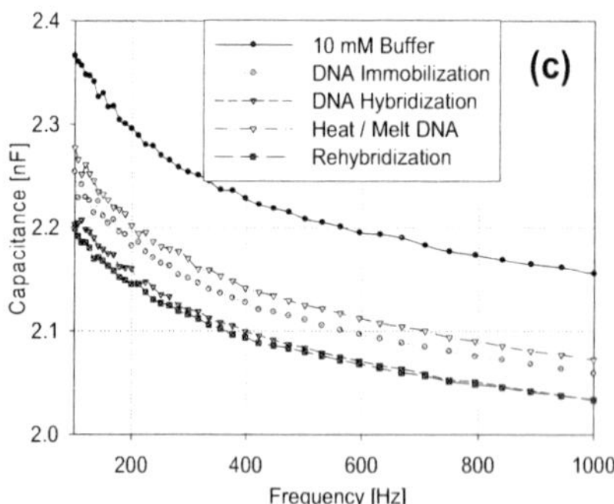

Fig 2. Capacitive detection of DNA immobilization, hybridization: (a) capacitance changes due to immobilization – changes are most significant below 10 kHz due to sensitivity of the double layer; (b) Electrode passivation using MCH – the passivated electrode does not show response to the complementary DNA, while the functionalized electrode shows a capacitance drop of about 3%. (c) thermal melting of hybridized DNA and DNA re-hybridization, which are clearly detectable events. (Buffer = 20 mM PB, Bias = 0.2V).

NANO2004-46052

Field Emission Study of Carbon Nanotubes: High Current Density from Nanotube Bundle Arrays

M.J. Bronikowski, H. M. Manohara, P.H. Siegel, B.D. Hunt

Jet Propulsion Laboratory, 4800 Oak Grove Drive, California Inst. of Technology, Pasadena, CA 91109

Abstract

We have investigated the field emission behavior of lithographically patterned bundles of multiwalled carbon nanotubes arranged in a variety of array geometries. Such arrays of nanotube bundles are found to perform significantly better in field emission than arrays of isolated nanotubes or dense, continuous mats of nanotubes, with the field emission performance depending on the bundle diameter and inter-bundle spacing. Arrays of 2-μm diameter nanotube bundles spaced 5 μm apart (edge-to-edge spacing) produced the largest emission densities, routinely giving 1.5 to 1.8 A/cm^2 at ~ 4 V/μm electric field, and >6 A/cm^2 at 20 V/μm.

Recent work [1-3] has shown that Carbon Nanotubes (CNTs) can have outstanding electrical field emission properties, with high emission currents at low electric field strengths (turn-on voltage as low as 1-3 V/μm and emission current as high as 0.1 mA from a single nanotube). [4,5] Carbon nanotubes are therefore attractive as cold-cathode field emission sources, especially for applications requiring high current densities (hundreds to thousands of amperes per cm^2) and lightweight packages, such as the recently proposed *nanoklystron*, [6,7] which is a micrometer dimension reflex klystron designed to generate milliwatts of power at terahertz frequencies. As part of our efforts to develop a high current density electron source for a working nanoklystron, we have investigated the field emission behavior of CNT arranged in a variety of geometries. We find that the best field emission behavior is achieved when the CNT are arranged in bundles a few microns in diameter, with the bundles arranged in arrays with an array spacing of several microns. CNT in this arrangement are found to have better field emission properties than either isolated, individual CNT or continuous, dense mats of CNT. We have studied the field emission characteristics of arrays of such CNT bundles, and have optimized field emission with respect to bundle size and separation.

CNT are produced by the decomposition of hydrocarbons over catalytic metal (iron) at elevated temperature. Substrates of Si or SiO2 are lithographically patterned with Fe (10 nm thick), then inserted into a tube furnace for CNT growth. Typical CNT growth conditions are: C_2H_4 flow, 380 sccm; H_2 flow, 190 sccm; total pressure, 200 Torr; Temperature, 650° C; growth time, 15 minutes. CNT grow upon the substrates only in the areas patterned with Fe catalyst. This catalyst was patterned in arrays of circular islands or dots with various diameter (in the range of 0.2 – 5.0 microns) and separation (in the range of 2 – 100 microns). Growth of CNT on the micron-sized dots of catalyst yields micron-sized zones of dense CNT having the appearance of ropes or bundles of nanotubes, separated by a distance equal to the separation of the catalyst dots before CNT growth; Figure 1 shows an example.

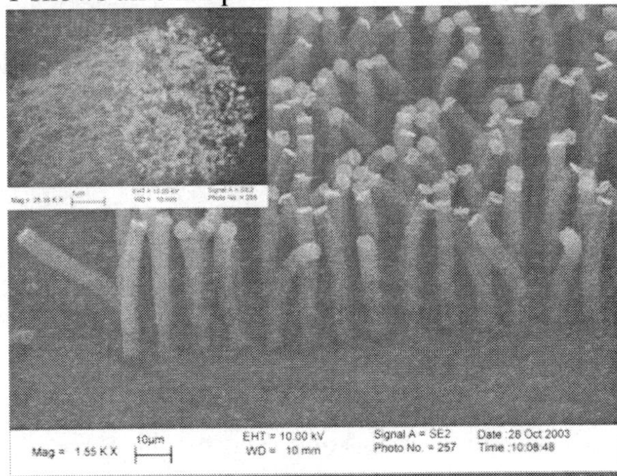

Fig 1. SEM image of CNT bundle array; inset shows one bundle with individual CNT visible.

For field emission measurements, CNT bundle arrays were grown in the pattern shown in figure 2. Dots with various diameters were written in arrays of size 0.5 mm × 20 mm. Six such arrays were written for each dot size, with six different edge-to-edge spacing between dots, as shown. The measurements were conducted in a diode mode using a tungsten probe anode of 100-μm tip diameter. The probe was scanned in Y-direction, and emission values were collected every 50 μm. (The X,Y scan coordinate convention is shown in Figure 2). Three lateral scans across the arrays were conducted at three different longitudinal locations (in X) separated by at least 1 mm. The measurement field was ~ 3 V/μm. The vacuum during measurement was maintained below 5×10^{-7} Torr.

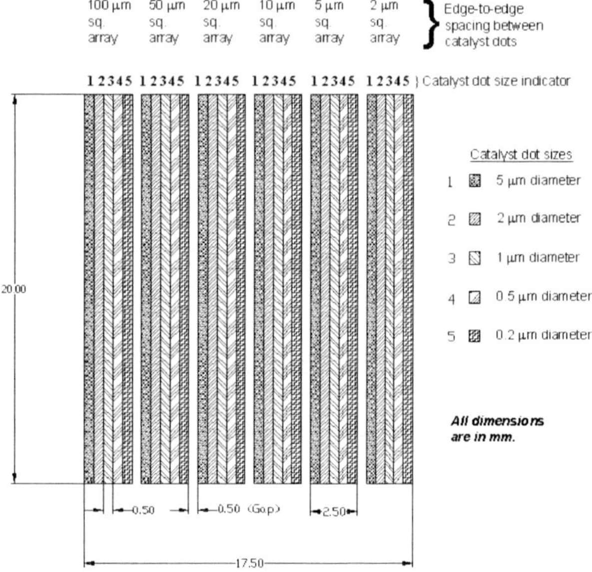

Fig 2. Schematic of an array layout of dots of iron catalyst from which CNT are grown.

The results of field emission current as a function of nanotube bundle diameters and inter-bundle spacing are shown in Figure 3. We observed substantial variation in emission with varying array spacing and the nanotube bundle diameter. Maximum emission was observed in the arrays with 5 μm edge-to-edge spacing, while almost no emission was observed for the arrays of spacing 50 μm and 100 μm.

The maximum emission current density observed was given by CNT bundles with diameter 1 – 2 microns and edge-to-edge bundle array spacing of 5 microns. Such bundle arrays routinely gave 1.5 to 1.8 A/cm^2 emission densities at ~ 4 V/μm field and >6 A/cm^2 at 20 V/μm. To the best of our knowledge, these are the highest current densities observed at such low fields. These observed current densities were much greater than the densities observed using either isolated single CNT (arranged in arrays with similar spacing) or dense mats of CNT grown from continuous Fe catalyst films.

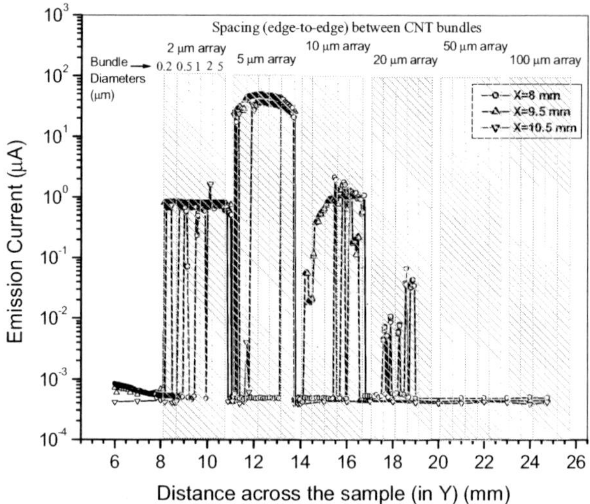

Fig 3. Field emission current from CNT bundle arrays as a function of array spacing and bundle diameter. The topmost labels refer to edge-to-edge CNT bundle spacing.

[1] A.A. Talin, K.A. Dean, and J.E. Jaskie, Solid-State Electronics, **45**, 963 (2001)

[2] J.M. Bonard., H. Kind, T. Stöckli, and L.-O. Nilsson, Solid-State Electronics, **45**, 893 (2001)

[3] H. Murakami, M. Hirakawa, C. Tanaka, and H. Yamakawa., Appl. Phys. Lett. **76**, 1776 (2000)

[4] W. Zhu, C. Bower, G.P. Kochankski, and S. Jin, Solid-State Electronics, **45**, 921 (2001)

[5] J-M. Bonard, H. Kind, T. Stockli, and L-O. Nilsson, Solid-State Electronics, **45**, 893 (2001)

[6] P.H. Siegel, T.H. Lee, J. Xu, JPL New Technology Report, NPO 21014, Mar. 21 (2000)

[7] P.H. Siegel, A. Fung, H.M. Manohara, J. Xu and B. Chang, Proc. 12th Int. Sym on Space THz Tech., 81 (2001).

NANO2004-46057

THE ROLE OF STM TIP SHAPE IN HEAT ASSISTED MAGNETIC PROBE RECORDING ON CONI/PT FILM

Li Zhang/DSSC CMU USA

James A. Bain/DSSC CMU USA

Jian-Gang Zhu/DSSC CMU USA

Leon Abelmann/Univ. Twente The Netherlands

Takahiro Onoue/Univ. Twente The Netherlands

ABSTRACT

A method of heat-assisted magnetic recording (HAMR) potentially suitable for probe-based storage systems is characterized. In this work, field emission current from a scanning tunneling microscope (STM) tip is used as the heating source. The tip is made of Ir/Pt alloy. Pulse voltages of 3-7 V with a duration of 500 ns were applied to a CoNi/Pt multilayered film. Written by a blunt tip (radius 1000 nm), marks are formed with a nearly uniform mark size of 170 nm when the pulse voltage is above 4 V. While sharp tip (radius 50 nm) writing achieves no mark. The emission area of our tip-sample system derived from an analytic expression for field emission current is approximately equal to the mark size, and is largely independent of pulse voltage. For the blunt tip, the emission region is almost the same as the mark size. While for the sharp tip, the initially formed mark is too small, so that the domain wall surface tension shrinks the mark and it crashes finally.

INTRODUCTION

The super-paramagnetic effect which induces the thermal relaxation of recorded information [1] is the fundamental obstacle to increasing magnetic recording density. To achieve thermal stability of recorded information, increases in the coercivity and anisotropy of the recording medium are needed. This makes traditional recording more difficult because conventional heads cannot generate sufficient field to switch the magnetization of the bits in thermally stable media. To overcome this obstacle, heat-assisted magnetic recording (HAMR), has been proposed [2]. HAMR draws on concepts from traditional magneto-optical (MO) recording for the writing process, but is not restricted to optical read-back.

In addition to optical heating methods suggested by extensions of MO recording, another possible approach to HAMR is the use of field emission current from a sharp metallic tip for heating. This has the possibility of very high spatial resolution as scanning tunneling microscopes (STM), which have similar architectures, show atomic resolution in surface observation [3-5]. Nakamura et al [6] demonstrated this writing method with an STM several years ago, and saw a mark size that increased with increasing tip voltage. We have also demonstrated the process previously [7-8], but saw very little dependence of mark size on tip voltage above a certain writing threshold. In this work, we complete the model of the writing process based on the Fowler-Nordheim theory [9] that quantitatively explains the effect of different tip radius on the mark size, and mark size dependence on pulse voltage.

EXPERIMENTS

The recording medium in these experiments is a CoNi/Pt multilayered film [8]. The same techniques were used to measure the perpendicular anisotropy, K_u, the saturation magnetization, M_S, and the coercivity, H_C, of the film at room temperature. We obtained K_u=2.5×10^5 J/m^3, M_S=3.4×10^5 A/m, and H_C=1.1×10^5 A/m. The exchange constant of the film, A, is estimated to be 1.0×10^{-12} J/m. Based on a theory of domain dynamics on the strong coupling perpendicular medium[7], the minimum stable domain size is calculated as 80 nm in this film. The Curie temperature of the film is about 250^{0}C.

A Digital Instruments Dimension 3000 scanning probe microscope (SPM), was used for writing and imaging. AFM (atomic force microscopy) mode was applied to scan the topographic features of the film, and MFM (magnetic force microscopy) mode was used to image the magnetic domain structures. STM was used for thermal writing. During STM scanning, the tip-sample junction is held at a bias voltage of 100 mV and set point current of 2 nA, from which the tip-sample spacing is estimated to be 0.7 nm during scanning [10]. The film is heated locally by applying pulses of 27 V in amplitude and 500 ns in duration, with a rise time of 100 ns to the sample. No externally applied field was used in these experiments, although it will clearly be necessary in any recording system implementation of this method. Demagnetizing fields from the surrounding region of the material switch the magnetization of the heated area.

RESULTS

Similar to our previous work, the written marks are magnetic in nature and the writing is reversible. The mark size is determined as the full-width half maximum of the MFM signal[8]. The dependence of mark size on the applied voltage and tip work function Fig. 1. For a blunt tip, the threshold voltage for writing is 4 V, and above threshold, the mark size is nearly uniform with an average 170 nm. For a sharp tip, no marks were seen after writing.

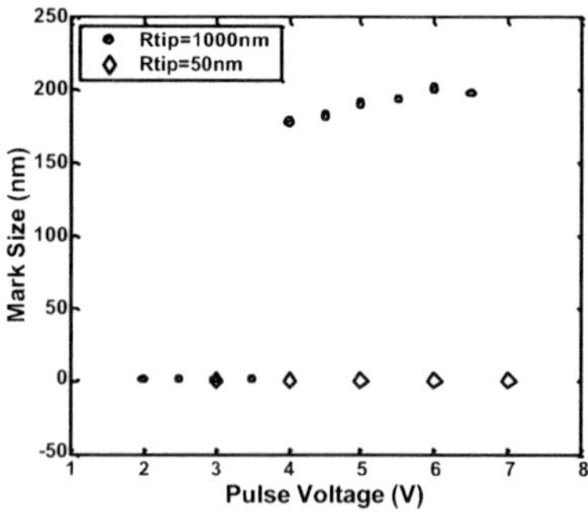

Fig. 1. Average mark size as a function of the applied pulse voltage, written by sharp and blunt tip, respectively.

A discussion of the field emission behavior is given below to explain these observations. In order to inform these discussions, scanning electron micrographs of both tips were taken and are shown in Fig. 2. We get that the radius for blunt tip is about 1000 nm, while 50 nm for sharp tip.

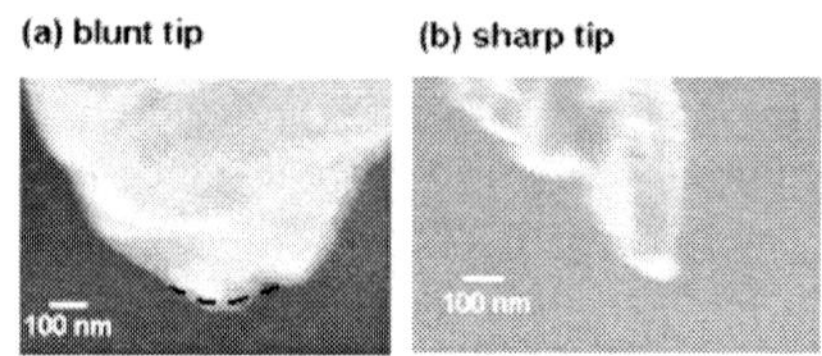

Fig. 2. SEM images for both tips.

DISCUSSION

From J. Simmons' theory [10], tunneling happens between a tip and surface when the bias voltage is below the work function of the tip. When the bias voltage exceeds the work function, field emission replaces the tunneling. The classical theory of field emission is based on the one-dimensional, planar Fowler-Nordheim equation [9]. Shown in Eq. (1) below:

$$J = \frac{AE^2}{\Phi} \cdot \exp\left\{ \frac{B}{\sqrt{\Phi}} - \frac{C\Phi^{\frac{3}{2}}}{E} \right\} \qquad (1)$$

The current density, J, is a function of the electrical field E between tip and sample, the work function of tip Φ. Other parameters, A, B, and C are shown in details in the work of Spindt et al [11]. An analytical extension of this theory to non-planar, tip-anode geometries was explored by Zuber et al [12]. Due to the shape of the tip, a prolate-spheroidal coordinate system is introduced [13], in which the orthogonal coordinates are (ξ, η, ϕ):

$$\begin{cases} x = a \cdot \sqrt{\xi^2 - 1} \cdot \sqrt{1 - \eta^2} \cdot \cos(\phi) & 0 \le \phi \le 2\pi \\ y = a \cdot \sqrt{\xi^2 - 1} \cdot \sqrt{1 - \eta^2} \cdot \sin(\phi) & 1 \le \xi \le +\infty \\ z = a \cdot \xi \cdot \eta & \eta_1 \le \eta \le \eta_2 \end{cases} \qquad (2)$$

In Eq. (2), a is the one-half distance between the hyperboli foci; ξ describes the radial distance along the curvature, η describes the sharpness of the tip, and ϕ is the angle around the central axis. In our problem, the tip is assumed to be axial-symmetric. So all the physical quantities are ϕ–independent. In the tip-sample system, the basic parameters are the tip radius r and the tip-sample spacing d. The half foci distance a is given as $a = d/\cos\theta$ where $\cos\theta$ is determined by r and d: $\tan\theta = \sqrt{r/d}$. Along the surface of tip, η is replaced by $\cos\theta$. The electrical field between the tip and anode is given as a function of ξ and bias voltage V [13]:

$$E(\xi) = \frac{V_0}{r} \cdot \frac{\tan\theta}{Q_0(\cos\theta) \cdot \sqrt{\xi^2 - \cos^2\theta}} \qquad (3)$$

From Eq. (1) we can get the current density. The edge of the emission current is defined as the point where $J(\xi_0)/J(1) = 0.001$, and we get:

$$\xi_0 = \sqrt{\left(\frac{3V\ln(10)}{B \cdot d \cdot \tan\theta \cdot Q_0(\cos\theta)} + \sin\theta \right)^2 + \cos^2\theta} \qquad (4)$$

The emission radius is expressed as:

$$R = d \cdot \sqrt{\sin\theta} \cdot \left[\frac{\xi_0 \sqrt{\xi^2 - \cos^2\theta} - \sin\theta}{\cos^2\theta} - \ln\left(\frac{\xi_0 + \sqrt{\xi_0^2 - \cos^2\theta}}{1 + \sin\theta} \right) \right]^{\frac{1}{2}} \qquad (5)$$

The emission current is calculated afterwards:

$$I(r, d, V, \Phi) = \int_0^{2\pi} d\phi \int_1^{\xi_0} d\xi \cdot J(\xi) h_\phi(\xi) h_\xi(\xi)$$

$$= \frac{2\pi \cdot A \cdot V^2}{\sin\theta \cdot Q_0(\cos\theta)} \cdot \int_1^{\xi_0} \frac{\exp\left[-\frac{Bd \cdot \tan\theta \cdot Q_0(\cos\theta)}{V} \cdot \sqrt{\xi^2 - \cos^2\theta} \right]}{\sqrt{\xi^2 - \cos^2\theta}} \cdot d\xi \qquad (6)$$

In our experiment, the tip-sample spacing d is fixed at 0.7 nm during pulsed writing. A plot of emission radius vs. tip radius based on Eq. (5) is shown in Fig. 3, with a fixed bias voltage 6 V. We observe that the emission radius increases with increasing tip radius r. Another important result is that the emission radius is not sensitive for the voltage in the range of 4-6 V. The I-V curve for both tips in our experiment is plotted in Fig. 4, based on Eq. (6). We observe a strong increase when the voltage is above the work function, which can be regarded as the threshold value.

42

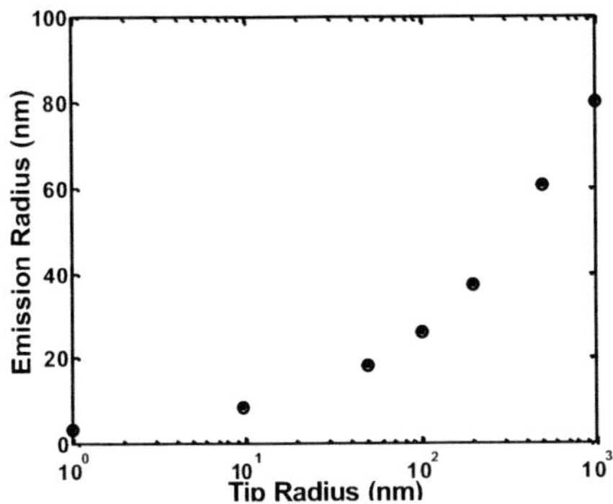

Fig. 3. Emission radius as a function of tip radius at 6V bias

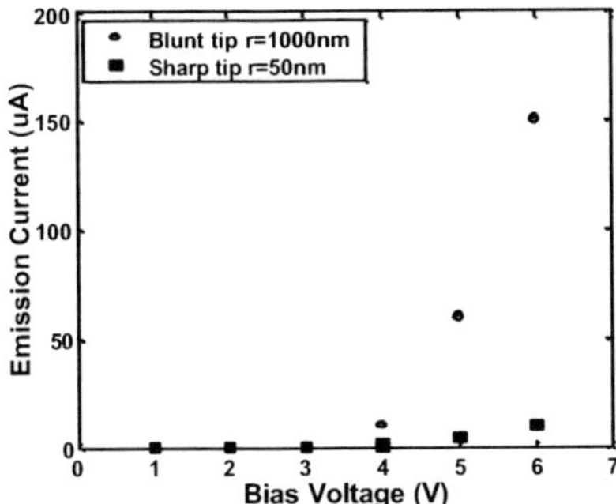

Fig. 4. Estimated I-V curve for both types of tips.

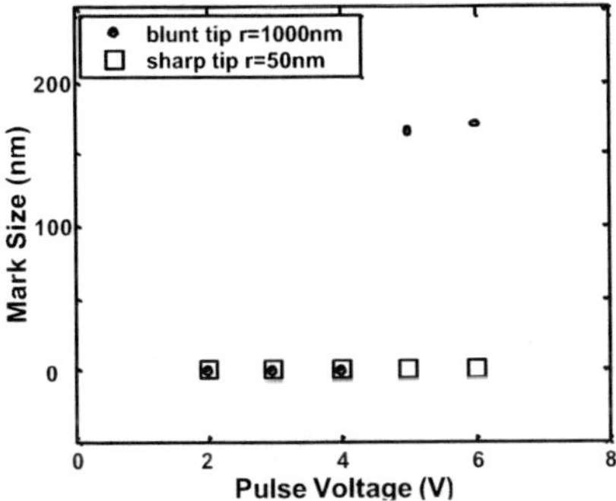

Fig. 5. Mark size extrapolated from the thermal profiles in the simulation by FEMLAB, for both tips in different voltages.

Mark size is determined by the thermal profile at the end of the heating cycle [7-8]. We have used FEMLAB to estimate the heated zone diameter. As a simple approximation, we assume the mark diameter is the same as the location of the thermal contour that is equal to 250^0C (the medium's Curie temperature). The model assumes the input power is the product of the current in Fig. 4 and the known applied voltage, and that the power is delivered uniformly over a circle with the same diameter as the beam. Fig. 5 shows a plot of the mark size estimated in this way from the FEMLAB simulations as a function of applied voltage for both tips. Note that this figure reproduces the threshold behavior and the weak dependence of mark diameter on applied voltage for a blunt tip, and null

writing by a sharp tip. For the blunt tip, no mark below threshold voltage is explained as the insufficient heating of the emission area. For the sharp tip, the heating by high pulse voltage is enough to locally raise the temperature above Curie point. However, due to the small size of the initially formed magnetic mark, the surface tension on the domain wall squeezes the mark till it crashes [7].

CONCLUSION

We have demonstrated a thermo-magnetic writing process using an STM on perpendicular CoNi/Pt multilayered medium. For the applied pulse voltage of 3-6 V, the written magnetic mark by a blunt tip (radius 1000 nm) has almost uniform size of 170 nm, and no marks were observed when written by a sharp tip (radius 50 nm). Mark size is close to the emission diameter (160 nm for the blunt tip) of our tip-sample system derived from a prolate-spheroidal coordinate system. Sharp tips result in small emission radius while recording medium being strong-coupling limits the writing of small marks. We propose to write stable small marks by using a sharper tip in a weak-coupling (or, more granular) film.

ACKNOWLEDGMENTS

This work is part of a project on MEMS based magnetic mass storage systems [14] at the Center for Highly Integrated Information Processing and Storage Systems (CHIPS) in Carnegie Mellon University. This work was funded in part by NASA grant # NAG8-1799.

REFERENCES

[1] P.L. Lu, and S.H. Charap, " High Density Magnetic Recording Media Design and Identification: Susceptibility to Thermal Decay" , IEEE Trans. Magn., Vol. 31, no. 6 (1995) pp. 2767-2769.

[2] J.J.M. Ruigrok, R. Coehoorn, S.R. Cumpson, and H.W. Kesteren, " Disk Recording beyond 100 Gb/in.^{2}: Hybrid Recording?" J. Appl. Phys., Vol. 87, no. 9 (2000) pp. 5398-5403.

[3] G. Binnig, H. Rohrer, Ch. Gerber, and W. Weibel, " Surface Studies by Scanning Tunneling Microscopy", Phys. Rev. Lett., vol. 49, no. 1 (1982) pp. 57-61.

[4] G. Binnig, H. Rohrer, Ch. Gerber, and W. Weibel, "Tunneling through a Controllable Vacuum Gap" , Appl. Phys. Lett., vol. 40, no. 2 (1982) pp. 178-180.

[5] G. Binnig, H. Rohrer, Ch. Gerber, and W. Weibel, "7×7 Reconstruction on Si (111) Resolved in Real Space" , Phys. Rev. Lett., vol. 50, no. 2, (1983) pp. 120-123.

[6] J. Nakamura, M. Miyamoto, S. Hosaka, and H. Koyanagi, " High-density Thermomagnetic Recording Method Using a Scanning Tunneling Microscope" , J. Appl. Phys., vol. 77, no. 2 (1995) pp. 779-781.

[7] L. Zhang, J.A. Bain, and J.-G. Zhu, " Dependence of Thermal-Magnetic Mark Size on Applied STM Voltage in Co/Pt Multilayers" , IEEE Trans. Magn. Vol. 38, no. 5 (2002) pp. 1895-1897.

[8] L. Zhang, J.A. Bain, J.-G. Zhu, L. Abelmann and T. Onoue, "A Model for Mark Size Dependence on Field

Emission Voltage in Heat Assisted Magnetic Probe Recording on CoNi/Pt Multilayers", to be published in IEEE Trans. Magn. (2004)

[9] R.H. Fowler and L.W. Nordheim, " Electron Emission in Intense Electric Fields", Proc. R. Soc. London Ser., A Vol. 119 (1928) pp. 173-181.

[10] J.G. Simmons, " Generalized Formula for the Electric Tunnel Effect between Similar Electrodes Separated by a Thin Insulating Film", J. Appl. Phys., 34 (1963) pp. 1793-1803.

[11] C.A. Spindt, I. Brodie, L, Humphrey, and E.R. Westerberg, " Physical properties of thin-film field emission cathodes with molybdenum cones", J. Appl. Phys., Vol. 47, no. 12 (1976) pp. 5248-5263.

[12] J.D. Zuber, K.L. Jensen, and T.E. Sullivan, "An Analytical Solution for Microtip Field Emission Current and Effective Emission Area", J. Appl. Phys., Vol. 91, no. 11 (2002) pp. 9379-9384.

[13] L.H. Pan, T.E. Sullivan, V.J. Peridier, P.H. Cutler, and N.M. Miskovsky, "Three-dimensional Electrostatic Potential, and Potential-energy Barrier, near a Tip-base Junction", Appl. Phys., Lett., Vol. 65, no. 17 (1994) pp. 2151-2153.

[14] L.R. Carley, J.A. Bain, G.K. Fetter, D.W. Greve, D.F. Guillou, M.S.L. Lu, T. Mukherjee, S. Santhanam, L. Abelmann, and S. Min, "Single-chip Computers with Microelectromechanical Systems-based Magnetic Memory", J. Appl. Phys., vol. 87, no. 9, (2000) pp. 6680-6685.

NANO2004-46073

Nanofluidic channel engineering using laminar flow layer-by-layer deposition of polyelectrolytes.

J. Collins[1], D. Goodman[1], P. Delhaes[2], A.P. Lee[1,3]
[1]Department of Biomedical Engineering, UC Irvine, USA, [2] Center Research Paul Pascal, Bordeaux, France
[3]Department of Mechanical and Aerospace Engineering, UC Irvine, USA

Layer-by-layer deposition of polyelectrolytes form multiple layers in the nanometer scale. Charged magnetic [1] and biomolecular [2] nano-colloids can be incorporated in to the layers. These multilayer assemblies are formed on the inner walls of a microchannel using laminar flow; alternatively polyanions and polycations are pumped through the channel. Water is also pumped through, between the deposition of each layer, in order to wash away the un-absorbed polyelectrolytes. The first charged layer is formed by silanation. The microfluidic device consists of a 'T' junction as shown in Fig. 1. Switching flow between polyelectrolytes and water is controlled by three syringe pumps, automated through a Labview virtual instrument.

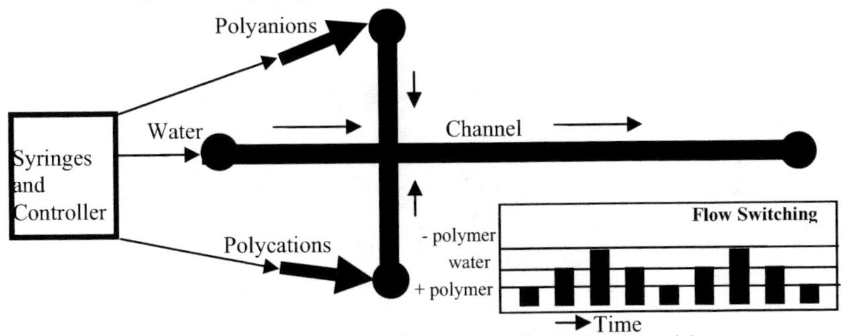

Fig 1. Experimental setup: Microfluidic layer-by-layer deposition

The thickness of the absorbed polyelectrolytes is observed 'in situ' using a pair of multiarray electrodes in the channel as shown in Fig. 2. In Fig 3, a preliminary experiment to detect the impedance spectra (Z) of flowing fluid is conducted. The three spectra shown are of nano-pure water, the polycation (poly(styrenesulfonate) or PSS), and polycation (Poly(allylamine hydrochloride) or PAH). The impedance change with the number of layers and the optimum flow rate for maximum deposition will be measured in the near future.

Because the impedance detection was not sufficiently accurate for a single channel, absorption thickness can also be measured using polyelectrolytes that exhibit electronic spectral peaks. Ferric hexacyanoferrate, or Prussian Blue, is one such polycation. By substituting Prussian Blue for PAH, and using UV absorption spectroscopy, the layers of PSS/PB is accurately detected (Fig. 4). Future experiments will attempt to identify a quantitative correlation between the number of absorbed layers and electrical impedance using a large array of serpentine channels and microarray electrodes.

Because polyelectrolyte solutions are miscible with each other, ample time must be allowed between flow switching to avoid mixing polycation and polyanion solutions within the channel.

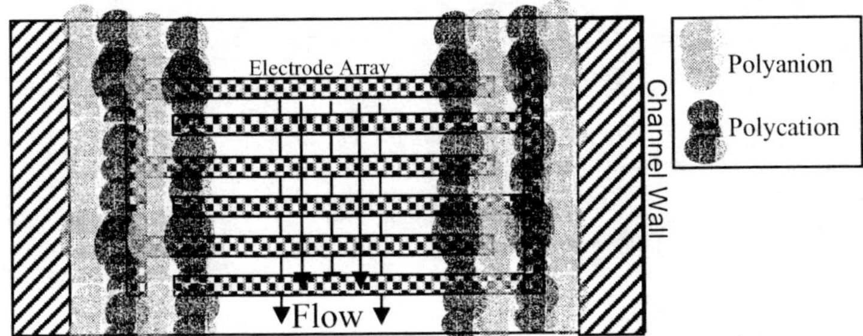

Fig 2. Top-Down view of the channel after layer-by-layer deposition. Microelectrode arrays can measure the relative absorption thickness.

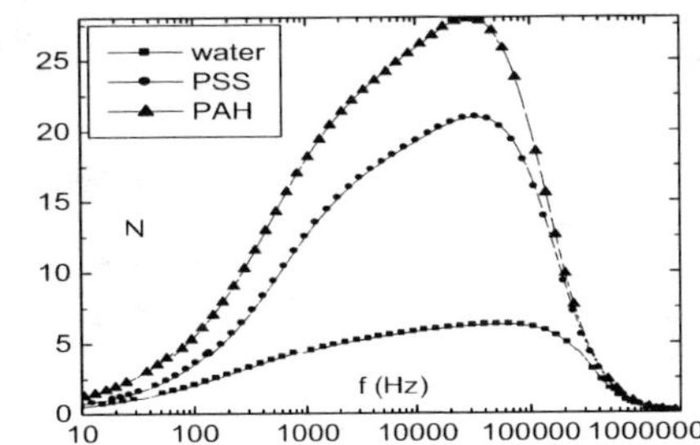

Fig 3. Impedance spectra of flowing fluids detected by microelectrodes in the channel.

By substituting dye for polyelectrolytes we predicted the extent of solution diffusion. Switching takes place between dye solution and water and by measuring the periodicity of the color change (Fig. 5), intervals between flow switching can be optimized and solution mixing can be effectively minimized. This experiment gives the minimum time required for the polyelectrolytes to fill the entire channel. In order to completely eliminate the polyelectrolyte diffusion a pulse of immiscible fluid (flourinert or oil) can be send between the aqueous fluids from a fourth inlet.

During flow switching, backflow from one channel inlet into an adjacent inlet became a problem. To overcome this, Tesla valves, shown in Fig. 6, were used. These structures, placed in sequence, effectively stopped the backflow and insured that the flow switching worked properly.

Thus nanoassembly of multilayers of polyelectrolytes can be coated in to the inner wall of micro or nano channels using electrostatic deposition. The mechanical stability and the effect of the surface modification will be studied in future.

Reference:

1. Jaiswal A, Collins J, Agricole B, Delhaes P, Ravaine S, Layer-by-layer self-assembly of Prussian blue colloids, J. Coll. Interface Sci., 261 (2): 330-335 (2003).
2. W. Tan, T. A. Desai, Layer-by-layer microfluidics for biomimetic three-dimensional structures, Biomaterials, 25 (2004) 1355-1364.

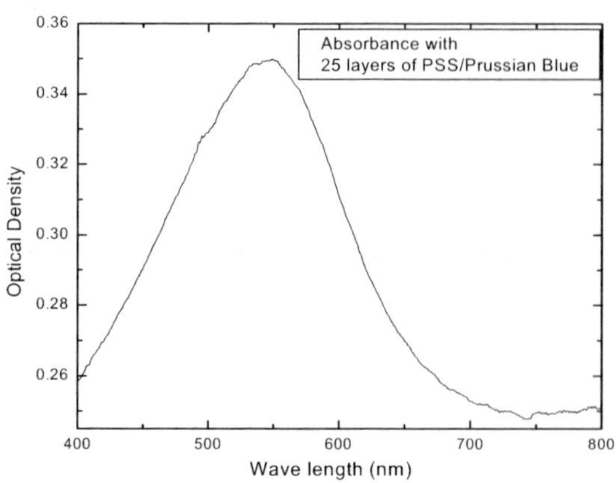

Fig 4. Optical Density spectra of PDMS after 25 alternated immersions in PSS/Prussian Blue

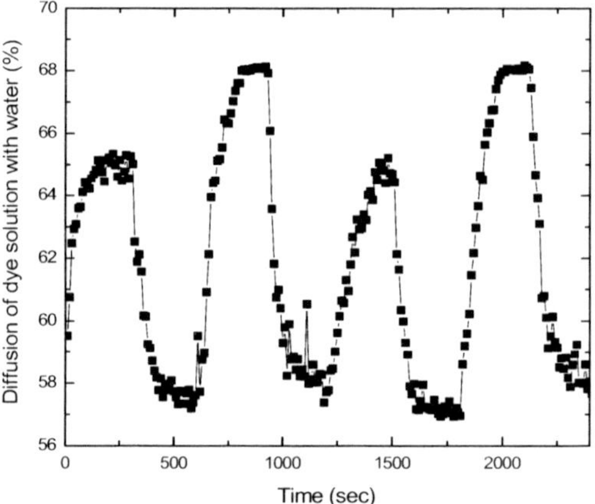

Fig 5. Time of flow of fluids is plotted against percent average pixel saturation, a measure of diffusion of dye solution with water.

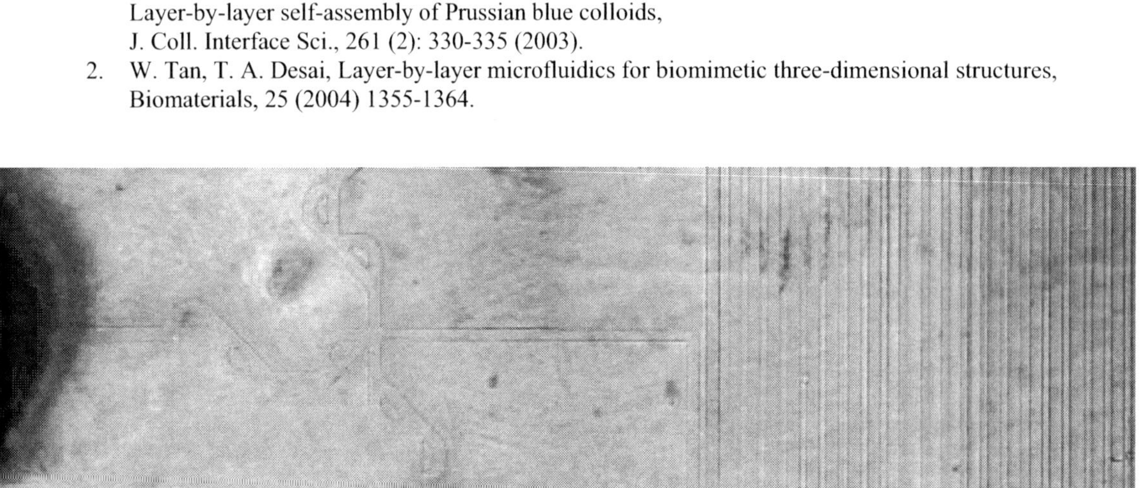

Fig 6. Optical Image of fluidic inlets with sequential Tesla valves to prevent backflow during fluidic switching.

NANO2004-46078

NANOTURF SURFACES FOR REDUCTION OF LIQUID FLOW DRAG IN MICROCHANNELS

Chang-Hwan Choi   Joonwon Kim   Chang-Jin "CJ" Kim

Mechanical and Aerospace Engineering Dept., University of California at Los Angeles (UCLA), CA 90095. USA

We report nano-engineered surfaces (NanoTurf), designed to make various micro- and nano-fluidic devices and systems less frictional for liquid flows, and describe microchannels made with such a surface. While our group has reported a dramatic (> 95%) drag reduction of discrete droplets flowing in a space between two parallel-plates covered with "random" nano-posts created by the "black silicon method" [1], this paper describes various nanofabrication techniques, including those capable of "designing" nanostructures with not only a good control of pattern sizes and periods but also practical manufacturability to be embedded in various micro- and nano-fluidic devices and systems. Microchannels are developed using the designed nanostructure surfaces and used for continuous flow tests.

Considerable drag reduction of liquid flow over a rough hydrophobic surface has been reported by several research groups [2-3]. But, most of the surfaces were obtained mainly by chemical treatment or modifying polymer surfaces. Our breakthrough is to fabricate surfaces that feature robust nano-size mechanical structures such as grates and posts, which can be controllable in size and period. Levitated by the hydrophobic nanostructures and flowing over air pocket between them, the liquid is expected to flow on NanoTurf with less friction than on flat surfaces (Fig. 1). It is critical that the surface structures are of nanometer scale, because otherwise the liquid would lose levitation as the pressure in liquid increases.

We have explored a series of techniques to fabricate NanoTurf. By using self-assembled synthetic materials (e.g., nanospheres and block copolymers) or organic materials (e.g., S-layer proteins) as the etching masks, we could transfer the nano-patterns onto silicon substrate. However, the control of pattern sizes and processing over a large area remain as the challenge. We also used the anodized thin-film aluminum as a nano-template, through which SiO_2 dots were deposited and used as the etch mask to transfer the nano-patterns onto silicon substrate after the lift off of the alumina layer (Fig. 2). Different sizes of alumina templates are possible depending on the process parameters in the anodization. Carbon nanostructures grown vertically by plasma enhanced hot filament chemical vapor deposition (PE-HF-CVD) were examined as well (Fig. 3). Nanoscale metal (e.g. Ni) dots are used as precursors for catalyst nanoparticles. Various pattern sizes and periods are possible, but it is very high temperature (~700 °C) process. Finally, the laser interference lithography was tested and favored for its superior control of nano-pattern sizes and periods as well as the large field area it can cover. The interference of two coherent laser beams to expose the resist creates an interference pattern of sinusoidal intensity (Fig. 4). Si nano-grating or nano-grid structures could be created by the interference lithography followed by the DRIE etching (Fig. 5). Regarding the hydrophobic coating, spin-coating of Teflon and conventional self-assembled monolayer coatings have been assessed.

Microchannels of various channel heights surfaced with the nano-grating (Fig. 5) or flat surfaces are fabricated, where it is challenging to protect the nanostructures (Fig. 6). The channel height is defined by the photoresist thickness and it is mechanically sealed. Experiments to measure the drag reduction of liquid flow in the microchannels are currently in progress with the high accuracy flow rate measurement system [4]. Phenomenal drag reduction (20~30%) has been obtained (Fig. 7), while the details are being reported elsewhere.

REFERENCES

[1] Kim, J. and Kim, C.-J., 2002, "Nanostructured surfaces for dramatic reduction of flow resistance in droplet-based microfluidics," Proc. IEEE MEMS 2002, pp. 479-482.
[2] Watanabe, K., Udagawa, Y. and Udagawa, H., 1999, "Drag reduction of Newtonian fluid in a circular pipe with highly water-repellent wall," J. Fluid Mech. **381**, pp. 225-238.
[3] Hasegawa, M., Nariai, H., Kaneko, K., Maki, H., Yabe, A. and Matsumoto, S. 1999, "Drag reduction on microscale concave-convex surface," ASME IMECE, pp. 317-322.
[4] U. Ulmanella, 2003, "Molecular Effects on the Boundary Condition in Micro and Nano Fluidic Channels," Ph.D. Dissertation, UCLA.

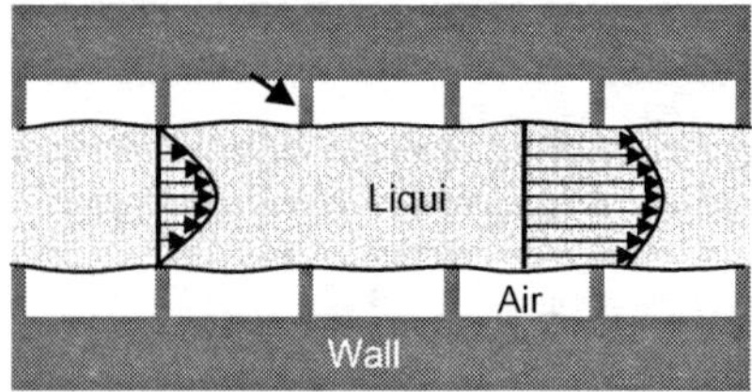

Figure 1. Concept of effective slip flow (drawn not to scale; nano-structures are much smaller in reality compared with the flow section). Liquid sits on hydrophobic nanostructures by surface tension. Majority of liquid boundary is with air where shear stress is negligible.

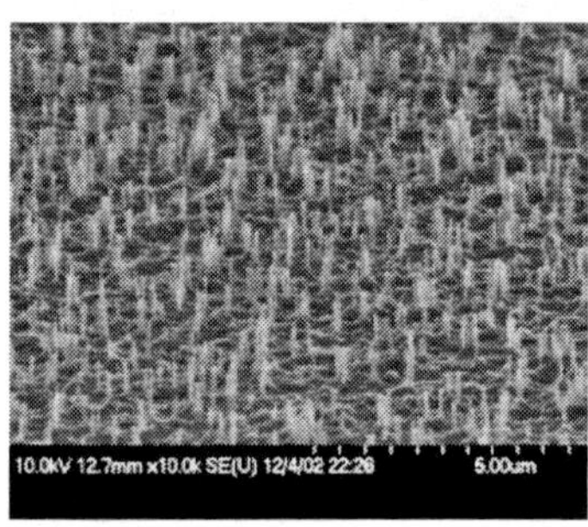

Figure 2. Si nanostructures made by using the anodized thin-film alumina as a nano-template

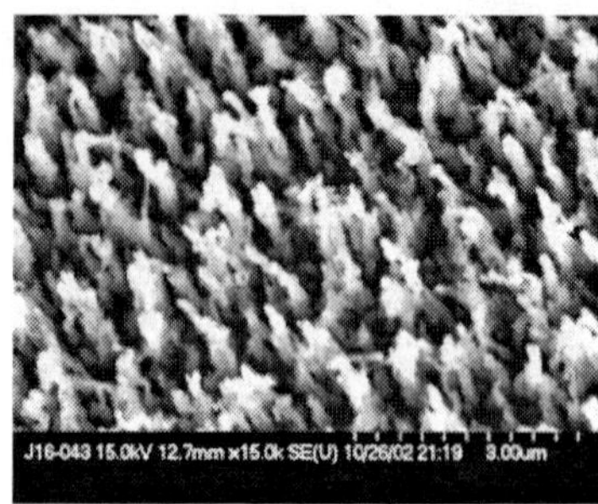

Figure 3. Carbon nanostructures (Samples from Dr. M. L. Simpson, Oak Ridge National Laboratory)

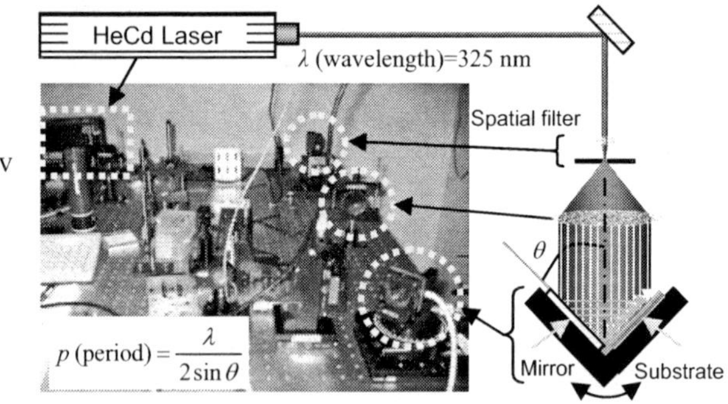

Figure 4. Picture and schematic of laser interference lithography system (Located at University of California at Santa Barbara)

$$p \text{ (period)} = \frac{\lambda}{2\sin\theta}$$

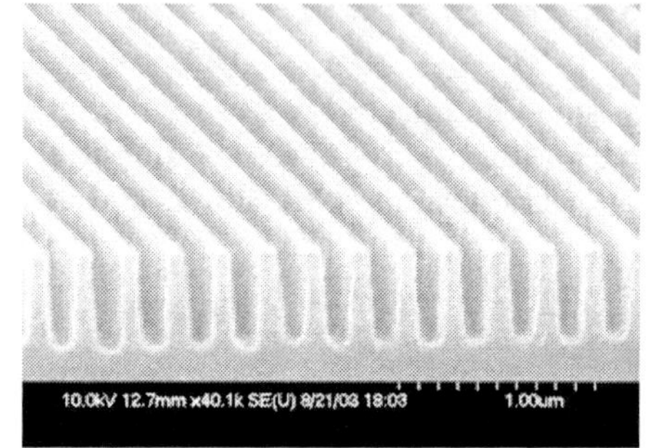

Figure 5. SEM image of Si nanograte structures made by interference lithography followed by DRIE etching and Teflon coating

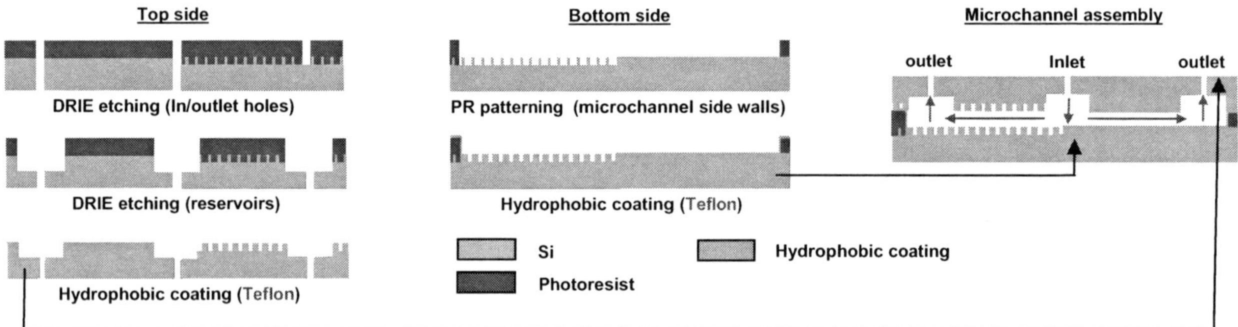

Figure 6. Procedure to construct microchannel for the flow experiment

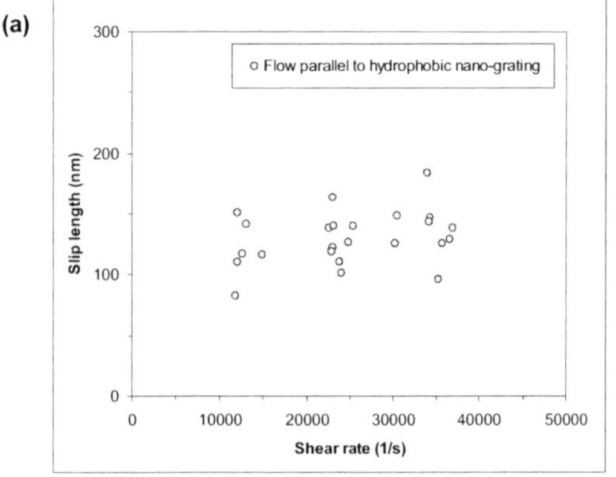

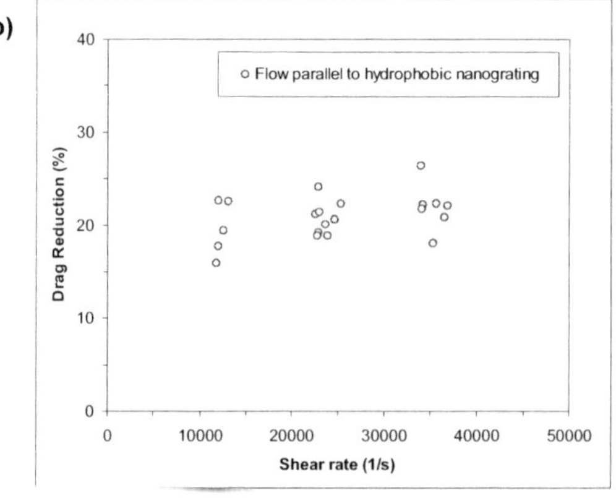

Figure 7. (a) Effective slip length of the hydrophobic nano-grating surface. (b) The corresponding drag reduction of water flow in a 3μm deep microchannel. Flow direction is parallel to the nano-grating direction.

NANO2004-46081

NONLINEAR COMPENSATOR DESIGN OF ACTIVE PROXIMITY SLIDERS FOR HEAD-DISK SPACING VARIATION SUPPRESSION IN 3-NM AIR BEARINGS

Jia-Yang Juang/ University of California, Berkeley

David B. Bogy/ University of California, Berkeley

ABSTRACT

Based on the concept that the FH of a portion of the slider that carries the read/write element can be adjusted by a piezoelectric actuator and the FH can be measured according to the magnetic signal, a new 3-DOF analytic model and an observer-based nonlinear compensator are proposed to achieve ultra-low FH with minimum modulation under short range attractive forces. Numerical simulations show that the FHM due to disk waviness is effectively reduced with a control voltage of less than ±2V.

INTRODUCTION

As the flying height (FH) is reduced to the 3-nm regime in ultra-high density hard disk systems, the flying height modulation (FHM) induced by the disk morphology and dynamic instability due to short range attractive forces become more significant. In order to achieve reliable reading and writing of magnetic data, it is required that the transducer location on the slider vibrates less than ±10% of the nominal FH, which means ±0.3 nm. Furthermore, considerable FHM may cause instability of the interface due to adhesive forces. The concept of FH adjustment by piezoelectric material has been proposed to decrease the effects of manufacturing tolerances and environmental variations on FH [1]. Li *et al* presented a real-time FH detection method by using readback signals [2]. Based on the concept that the FH of a portion of the slider that carries the read/write element can be adjusted by a piezoelectric actuator in between the slider and suspension and the FH can be measured according to magnetic signal, a new analytical model and design of a robust sliding mode control are proposed for suppressing the FHM of ultra-low flying height air-bearing sliders including the effects of short range attractive forces and disk morphology. The attractive forces are included in the model as a highly nonlinear term and the effect of disk morphology is modeled as unknown but bounded disturbances.

NONLINEAR 3-DOF LUMPED MODEL

The cantilever, composed of a piece of piezoelectric material and a portion of the slider, deflects under an electric voltage V and an external force F vertically exerted on the tip.

The constitutive equation of the tip deflection subject to a voltage and a force can be described as follows [3]:

$$\delta = aF + bV \tag{1}$$

$$a = \frac{1}{k} = \frac{4L_c^3}{E_p w t_p^3} \frac{\alpha\beta(1+\beta)}{\alpha^2\beta^4 + 2\alpha(2\beta + 3\beta^2 + 2\beta^3) + 1}$$

$$b = \frac{3L_c^2}{t_p^2} \frac{\alpha\beta(1+\beta)}{\alpha^2\beta^4 + 2\alpha(2\beta + 3\beta^2 + 2\beta^3) + 1} d_{31}$$

$$\alpha = E_c / E_p, \quad \beta = t_c / t_p$$

According to Eq. (1), the cantilever is modeled as a single DOF mass-damper-spring system. With the assumption that the ABS design is symmetric and the flying condition is at 0° skew, the motion in the roll direction has little contribution to the system response. However, the two pitch modes will contribute to the slider's dynamics at the R/W transducer. The active proximity is then modeled as a 3-DOF system as shown in Fig. 1. Note that there are two nonlinear elements. One is the nonlinear air bearing spring at the trailing edge,

$$k_1(FH_{pt}) = \beta' \cdot (FH_{pt})^{\alpha'} = 5.1 \cdot (FH_{pt})^{-0.48} \tag{2}$$

The other is the short range attractive force acting on the cantilever tip, which we model here as intermolecular van der Waals forces

$$F_{imf} = -A' / (FH_{pt})^3 + B' / (FH_{pt})^9 \tag{3}$$

where the constants α' and β' depend on the ABS design while A' and B' are related to the material properties of slider and disk. In this paper, the values in [4] are used.

The equation of motion of this model is written as follows:

$$[m]\{\ddot{x}\} + [c]\{\dot{x}\} + [k]\{x\} = \{F\} \tag{4}$$

where

$$[m] = \begin{bmatrix} M & 0 & 0 \\ 0 & I_\theta & 0 \\ 0 & 0 & m \end{bmatrix} ; \quad [c] = \begin{bmatrix} c_2 + c_3 & -l(c_3 - c_2) & -c_3 \\ -l(c_3 - c_2) & l^2(c_2 + c_3) & lc_3 \\ -c_3 & lc_3 & c_1 + c_3 \end{bmatrix}$$

$$[k] = \begin{bmatrix} k_2 + k_3 & -l(k_3 - k_2) & -k_3 \\ -l(k_3 - k_2) & l^2(k_2 + k_3) & lk_3 \\ -k_3 & lk_3 & k_1 + k_3 \end{bmatrix}$$

$$\{x\} = \begin{bmatrix} z_2 \\ \theta \\ z_1 \end{bmatrix}; \quad \{F\} = \begin{bmatrix} f_{d2} \\ lf_{d2} \\ F_{imf} - u + f_{d1} \end{bmatrix} = \begin{bmatrix} k_2 d_2 + c_2 \dot{d}_2 \\ l(k_2 d_2 + c_2 \dot{d}_2) \\ F_{imf} - u + (k_1 d_1 + c_1 \dot{d}_1) \end{bmatrix}$$

Note that the disk profile d_1 and d_2 are assumed unknown but bounded.

Numerical simulations were used to calculate the responses of the system to disks of various waviness wavelengths. The peak-to-peak amplitude of the waviness was 2 nm, and the wavelengths were chosen between 2.5 and 0.02 mm. The FHM is obtained by subtracting d_1 from z_1. The result is shown in Fig. 2. It is observed that FHM is relatively small except for the wavelength of about 0.1 mm, corresponding to a frequency of 150 kHz for a radius of 20 mm and disk RPM of 7200.

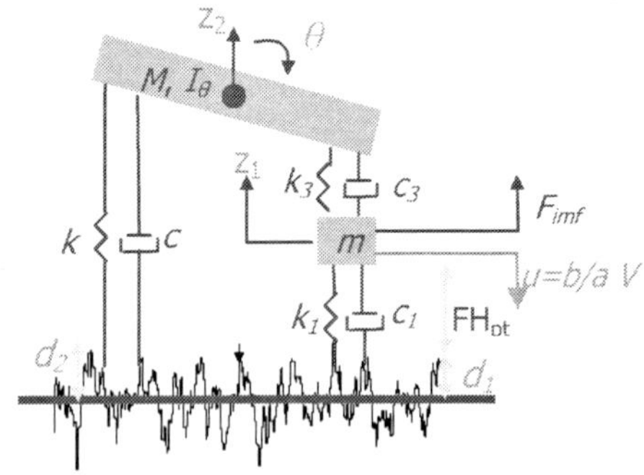

Fig. 1. Schematic diagram of 3-DOF dynamic model of active proximity sliders.

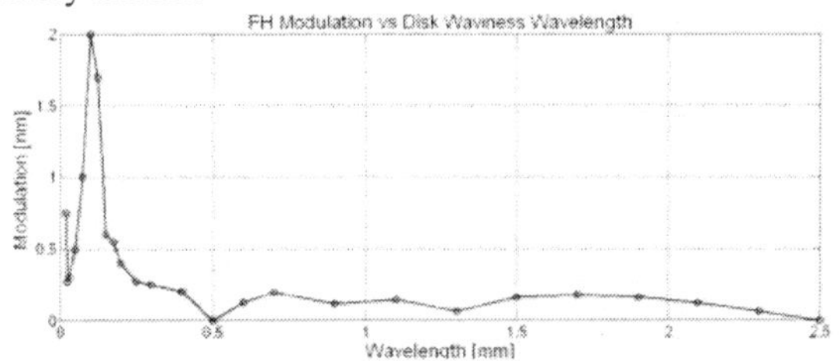

Fig. 2. Peak-to-peak FHM vs waviness wavelength for the active proximity slider without control.

DESIGN OF NONLINEAR COMPENSATORS

The FHM due to some particular disk wavelengths are as large as 2 nm. It is desirable to suppress this modulation by feedback control. Because of the nonlinear components and uncertain disturbance in the air bearing systems, an observer-based nonlinear control or nonlinear compensator design approach is used. Assuming that the real-time FH or magnetic spacing can be measured, we first built an observer for the state estimation and designed a sliding control law using the observer as the plant.

Equation (4) is transformed to a state-space representation as follows:

$$\begin{cases} \dot{x} = Ax + Bu + f(x) + f_d \\ y = Cx \end{cases} \tag{5}$$

And the observer is

$$\dot{\hat{x}} = A\hat{x} + Bu + f(\hat{x}) + L(y - C\hat{x}) \tag{6}$$

The error dynamic is shown as follows

$$\tilde{\dot{x}} = \dot{\hat{x}} - \dot{x} = (A - LC)\tilde{x} \quad f_d \tag{7}$$

where $f(.)$ and f_d represent the nonlinear components and disturbances, respectively. The control goal is to push the FHM

to zero. If z_1 is used as a state, this will be a tracking problem, $z_1 \rightarrow d_1$. However, the future d_1 is unknown. In order to resolve this, a new state $z_3 = z_1 - d_1$ is used. The states of the system are $x = [z_2 \ \dot{z}_2 \ \theta \ \dot{\theta} \ z_3 \ \dot{z}_3]^T$. d_1 is then regarded as disturbances to the system. The observer gain matrix $\boldsymbol{L}$ is chosen as in a Luenberger observer so as to place the poles of ($\boldsymbol{A\text{-}LC}$) at desired locations.

The sliding surface is defined as

$$s = \hat{x}_6 + \lambda \hat{x}_5$$

$$\begin{aligned} \dot{s} &= \dot{\hat{x}}_6 + \lambda \dot{\hat{x}}_5 \\ &= \frac{1}{m}[k_3 \hat{x}_1 + c_3 \hat{x}_2 - lk_3 \hat{x}_3 - lc_3 \hat{x}_4 - k_3 \hat{x}_5 - (c_1 + c_3)\hat{x}_6 + \\ &\quad F_{imf} - k_1 \cdot x_5 - u] + L_6(x_5 - \hat{x}_5) + \lambda[\hat{x}_6 + L_5(x_5 - \hat{x}_5)] \end{aligned} \tag{8}$$

The control law is designed as

$$\begin{aligned} u &= k_3 \hat{x}_1 + c_3 \hat{x}_2 - lk_3 \hat{x}_3 - lc_3 \hat{x}_4 - k_3 \hat{x}_5 - (c_1 + c_3)\hat{x}_6 + \\ &\quad F_{imf} - k_1 \cdot x_5 + m\lambda[\hat{x}_6 + L_5(x_5 - \hat{x}_5)] + mL_6(x_5 - \hat{x}_5) + m\eta s \end{aligned} \tag{9}$$

such that

$$\dot{s}s = -\eta s^2 \tag{10}$$

To investigate the controller performance, a numerical simulation was conducted at the disk wavelength of 0.1 mm. The FHM is reduced almost to zero with a driving voltage less than $\pm 2V$ as shown in Fig. 3.

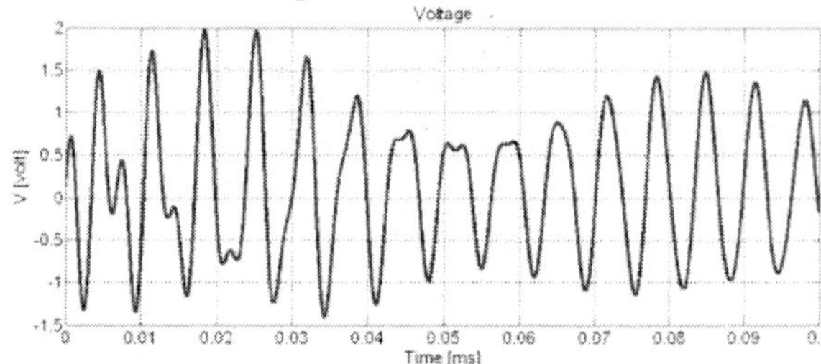

Fig. 3. The voltage determined by the control law

CONCLUSIONS

A new 3-DOF dynamic model is proposed to model active proximity sliders, which are actuated by a piece of piezoelectric material. An observer-based nonlinear compensator is designed based on the model. Numerical studies show that the FHM due to disk waviness is effectively reduced.

REFERENCES

[1] Kurita, M. and Suzuki, K., 2004, "Flying-Height Adjustment Technologies of Magnetic Head Sliders," *IEEE Trans. Magn.*, vol. 40, no. 1, pp. 332-336.

[2] Li, Amei., Liu, Xinqun., Clegg, W., Jenkins, D. F. L., and Donnelly, T., 2003, "Real-Time Method to Measure Head Disk Spacing Variation Under Vibration Conditions," *IEEE Trans. Instru. And Measur.*, vol. 52, no. 3, pp. 916-920.

[3] Smits, J. G. and Choi, W.-S., 1991, "The Constituent Equations of Piezoelectric Heterogeneous Bimorphs," *IEEE Trans. Ultrason., Ferroelect., Freq. Contr.*, vol. 38, no. 3, pp. 256-270.

[4] Thornton, B. H., 2003, "Head-Disk Interface Dynamics of Ultra-Low Air Bearing Sliders for Hard-Disk Drive Applications," *Ph D dissertation*, University of California, Berkeley.

NANO2004-46086

Micromachined Ultrasonic ElectroSpray Source Array for High Throughput Mass Spectrometry

Samuel Aderogba, J. Mark Meacham, F. Levent Degertekin, and Andrei G. Fedorov[*]
George W. Woodruff School of Mechanical Engineering (ME)
[*]Corresponding Author: 404-385-1356 (phone) & E-mail: andrei.fedorov@me.gatech.edu

Introduction

According to the recent Laboratory News' Proteomics Special article[1] Mass Spectroscopy (MS) has become the technology of choice to meet today's unprecedented demand for accurate bioanalytical measurements, including protein identification. Although MS can be used to analyze any biological sample, it must be first converted to gas-phase ions before it can be introduced into a mass spectrometer for analysis. It is transfer of a very small liquid sample (proteins are very expensive and often very difficult to produce in sizable quantities) into a gas-phase ions that is currently considered to be a bottleneck to high throughput proteomics[2]. Electrospray ionization (ESI) is a technique developed in early 1990th to generate a spray gas-phase ions by applying high voltage (from several hundreds volts and up to a few thousands kilovolts relative to the ground electrode of the MS interface) to a small capillary through which the liquid solution is pumped[3]. The high electric field ionizes the fluid forming the converging Taylor cone of the exiting jet which eventually breaks into many small droplets when the repulsive Coulombic forces overcome the surface tension. Because of the focusing effect associated with the spraying the electrically charged fluid, the size of the electrospray cone and thus of the formed droplets is in a few tens of nanometers range although the inner diameter of the capillary is in the micrometer range.

Recently, significant efforts went into development of the silicon-based micro/nano chip interface for producing electrospray. To the best of our knowledge, the proposed approaches are still based on using a charged capillary to create electrospray, as has it has been done traditionally, with innovation coming from using batch MEMS fabrication to reduce the cost and improve yield[4,5,6]. In the last year, we have been developing a conceptually new, drop-on-demand electrospray chip that is using the ultrasonic waves to drive the flow and capable of operating with the smallest reagent samples, resulting in the highest sensitivity and potentially requiring much lower voltages for efficient ionization[7]. The proposed electrospray generation method is utilizing the MEMS based ultrasonic aerosolizing technology[8] that is low-cost, disposable, and able to produce charged liquid droplets of size and uniformity required for effective protein analysis. Further, since Taylor cone formation is not required for atomization in the proposed device, potentially much lower operating voltages would be needed for ion formation, leading to much gentler atomization process and reduction in molecule fragmentation. In contrast to the conventional electrospray implementation, the proposed technology allows for further, highly controlled nozzle miniaturization, multiplexing or a parallel analysis of many samples simultaneously using an array of nozzles, it reduces the distance between the electrospray interface and MS reducing the sample exposure and loss, and the chips are made disposable eliminating the sample-to-sample contamination and solvent and time consuming chip-washing step. In addition, this approach should allow to expand a set of possible solvents that can be used for MS.

Micromachined Ultrasonic Electrospray Source Array

We have developed a first prototype of the proposed device and successfully tested it with various electrolytes and dielectric fluids to obtain preliminary proof-of-concept results. The schematic of the prototype device, which uses a PZT-5H transducer and a silicon micromachined reservoir and nozzle is shown in Fig. 1-a. The nozzle and the focusing horn are formed by a single step wet etching a silicon wafer in KOH. This single, one mask MEMS process results in both low cost fabrication and a unique acoustic wave focusing structure enabling atomization at low pressure, temperature, and operating power. The scanning electron micrograph of the fabricated silicon nozzles is shown in Fig. 1-b. Finite element simulations of the pressure distribution generated by the acoustic waves in the nozzle also show the expected focusing effect (Fig. 1-c).

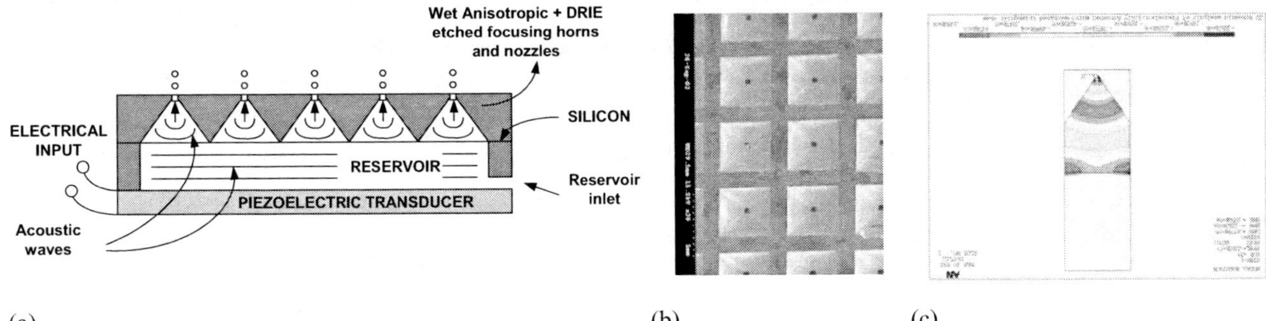

(a)   (b)   (c)

Figure 1. a) Schematic of the prototype droplet generator. b) SEM of the KOH etched silicon nozzles. c) The finite element simulation of focused acoustic pressure distribution in the nozz

We have used this prototype device to eject both water-methanol solution (electrolyte) and high purity DI water (dielectric) to prove that we can ionize electrolytes at low applied voltages. Figure 2-a shows the picture of the device in operation where the clouds of aerosol generated is seen above the device. Figure 2-b and 2-c shows enhanced view of droplets ejected from several nozzles simultaneously and a stroboscopic image of a single jet of 5 um droplets, respectively.

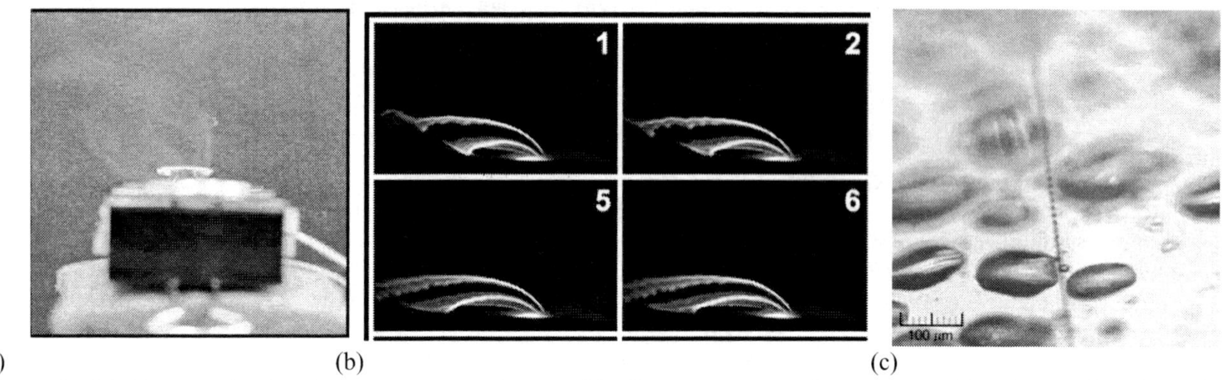

(a)   (b)   (c)

Figure 2. (a) Droplet ejection from a prototype device. (b) Close-up of several nozzles working in parallel (c) A jet of individual 5 um droplets ejected by a device.

Figure 3 shows the critical proof-of-concept experiment that demonstrates the possibility of generation of charges droplets with the applied DC voltage of only few tens of Volts rather than kilovolts that are required by conventional electrospray sources. Figure 4a shows the current (converted into a plotted voltage signal) that is detected by the current collector (metal plate) upon impingement of the charged droplets ejected from the atomizer array. Clearly, even when only 100 volts is applied, there is significant electrical charging of the droplets, resulting in an increase in the collected current. To make sure that observed results are not artifacts, the negative control experiment with a dielectric fluid (high purity DI water) has also been performed. As clearly seen on Fig. 3b, no ionization occurs when dielectric fluid is ejected even at high applied voltages. The non-zero baseline current (voltage) shown in both Figs. 3a and 3b are due to electromagnetically induced noise in the current/voltage amplifier arising from the voltage source which is used to drive the ejector array.

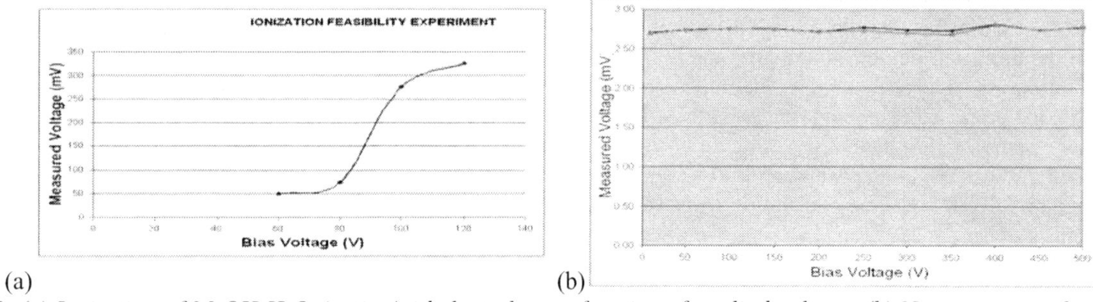

(a)   (b)

Figure 3. (a) Ionization of MeOH:H_2O:Acetic Acid electrolyte as function of applied voltage. (b) Negative control: absence of ionization when dielectric fluid is utilized regardless of the applied bias voltage.

Acknowledgments. We would like to acknowledge insightful discussions and advice from Professor Facundo Fernandez (Bioanalytical Mass Spectrometry, Georgia Tech Chemistry Department).

[1] "Automated NanoElectrospray: A New Advance for Proteomics Researchers", LABORATORY NEWS, 2002

[2] Hager, D. B., Dovichi, N. J., Klassen, J. and Kebarle, P. (1994), "Droplet electrospray mass spectrometry," *Analytical Chemistry*, **66** (22): 3944-3949.

[3] Gaskell, S. J. (1997), "Electrospray: principles and practice," *Journal of Mass Spectrometry*, **32**: 677-688.

[4] Desai, A., Tai, Y.-C., Davis, M. T. and Lee, T. D. (1997), "A MEMS electrospray nozzle for mass spectrometry," *1997 International Conference on Solid State Sensors and Actuators*, Chicago, Il, pp. 927-930.

[5] Li, J., Thibault, P., Bings, N. H., Skinner, C. D., Wang, C., Colyer, C. L. and Harrison, J. (1999), "Integration of microfabricated devices to capillary electrophoresis-electrospray mass spectrometry using a low dead volume connection: application to rapid analysis of proteolytic digests," *Analytical Chemistry*, **71**: 3036-3045.

[6] Licklider, L., Wang, X.-Q., Desai, A., Tai, Y.-C. and Lee, T. D. (2000), "A micromachined chip-based electrospray source for mass spectrometry," *Analytical Chemistry*, **72**: 367-375.

[7] Fedorov, A. G. and F. L. Degertekin, (2003), "Micromachined Ultrasonic Electrospray Source Array for Mass Spectrometry", *US Patent Pending*.

[8] Meacham, J. M., Ejimofor, C., Kumar, S., Degertekin, F. L. and Fedorov, A. G. (2004), "A micromachined ultrasonic droplet generator based on a liquid horn structure," *Review of Scientific Instruments*, **75** (5), 1347-1352.

NANO2004-46094

THE ENGINES OF BIOMOLECULAR MOTORS

Jung-Chi Liao, George Oster
Department of Molecular and Cell Biology and ESPM, University of California, Berkeley, CA 94720

INTRODUCTION

The majority of biomolecular motors are powered by nucleoside triphosphate (NTP), especially adenosine triphosphate (ATP). These motors consist of a β-sheet with highly conserved motifs and the nucleotide binding domain around it. The highly conserved protein folds are the *engines* of these motors, which convert the energy of NTP hydrolysis cycle to mechanical work. Although functions of molecular motors are widely diverse, (including cargo movement, DNA unwinding, protein degradation, ion pumping, etc), the nucleotide binding domains are very similar. In the binding site, NTP undergoes a hydrolysis cycle

$$E + NTP \rightleftharpoons E \cdot NTP \rightleftharpoons E \bullet NTP \rightleftharpoons E \bullet NDP \bullet P_i \rightleftharpoons E \bullet NDP + P_i \rightleftharpoons E + NDP + P_i$$

where E is the enzyme (motor protein), the small dot represents the docking of NTP, and the large dot represents the tightly-bound states. The hydrogen bond network formed in the NTP binding step, as shown in Figure 1 [1], deforms the β-sheet and adjacent structures. The local deformation propagates to conformational changes of functional residues to do mechanical work or to change the affinity to the substrate [2]. For multimeric motor proteins, we must also consider the stress paths among subunits which control the sequence and the activity of the protein. Stress trajectories emanating from a binding site either passes through a circumferential stress loop or a stress loop through the substrate.

Here we describe the mechanisms of T7 hexameric helicase, F_1 ATP synthase, and φ29 portal protein to illustrate the similarity and the difference of the mechanochemical couplings.

T7 HELICASE

Bacteriophage T7 helicase is a hexameric motor protein that couples energy from dTTP (deoxythymidine triphosphate) hydrolysis cycle to unwind double-stranded DNA by translocating along one DNA strand [3]. Based on the crystal structure [4] and the biochemical measurements [3, 5], we propose a 6-state, 6-subunit, sequential kinetic-network model.

The catalytic sites communicate via mechanical strain. The central player in this inter-site communication is the β-sheet that lies close to each catalytic site. The closing of the catalytic site around the nucleotide deforms the ß-sheet, and this deformation moves the contacting DNA. The elastic energy stored in this sheet as it deformed will power the recovery stroke that resets the power loop after it detaches from the DNA. At the same time, a third ß-sheet loop transfers strain to the P-loop of the next catalytic site. The data for the transient kinetics and the simulation results based on the mechanical consideration are shown in Figure 2A.

F_1 ATP SYNTHASE

F_1 ATP synthase is a rotary motor protein which can either manufacture ATP or hydrolyze ATP to pump ions [1, 2]. This motor is composed of a hexamer and a central rotating γ-subunit. The hexamer contains 3 α-subunits and 3 β-subunits. Only 3 β-subunits are catalytic. The rotation of the γ-subunit was observed in the hydrolysis direction by attaching an actin filament to this subunit [6]. We developed a mechanochemical model to explain the experimental results [2, 7].

In the catalytic site, the sequential hydrogen bond formation during ATP binding is converted to elastic strain energy of ~24 kT. This strain energy bends the β-subunit, which is an elastic element to store some of the energy for later use [2, 7, 8]. The bending stress is propagated to the γ-subunit as a rotary torque. Each of the subunits cycle through its conformational range in sequence, and all 3 β-subunits provide a continuous rotational torque to rotate the γ-subunit by 360°. Stresses propagate through at least two switch points on γ-subunit that interact with specific sites of the hexamer, and also communicate to other catalytic sites through circumferential contacts. We thus build the potential energy term by term by including elastic energy of the subunits and electrostatic energy of interactions. In order to reach the observed high efficiency, each subunit must exert two power strokes in two different steps: ATP binding and Pi release. The energy of the second step comes from the stored energy when ATP binds. This model quantitatively explains major features of the F_1 motor.

φ29 PORTAL PROTEIN

Bacteriophage φ29 portal protein is a hexameric or a pentameric ATPase which pushes DNA into the phage's capsid against high pressure [9]. The sequence alignment against other NTPases shows that it is likely that the β-sheet and the ATP contacting residues are conserved. We thus assume the coupling between the hydrogen bond network and the conformational changes is similar to other NTPases we have studied. We develop a mechanochemical model to explain the observation in the pulling experiments [9].

According to the experimental results, we propose two possible DNA translocation mechanisms for the portal protein. The mechanical step can be either coupled to the ATP binding step or the ADP release step. In the first model, the ATP-bound state has high affinity to DNA, so that the hydrogen bond network formed in the ATP binding step propagates stress to move the DNA contacting residues. The energy of hydrogen bond formation is directly transduced to the mechanical work of DNA movement. In the other model that ADP release is coupled to the mechanical step, the ADP-bound state has high affinity to DNA, so that DNA can be moved only in the ADP-bound state. The energy of hydrogen bond formation during ATP binding is stored as elastic energy, and the energy is released when the site is recoiled to its original conformation during the ADP release process. A possible experiment to distinguish these two models is to conduct a bulk experiment to test the DNA affinity to the enzyme in both ATP abundant and ADP abundant environments. Figure 2B shows the simulation results for different external pulling forces under various [ATP] based on the ATP binding model. The simulation is in good agreement with the experimental data.

CONCLUSIONS

In all three biomolecular motors, NTP binding forms the hydrogen bond network to deform the structures around the conserved β-sheet. The stresses initiated at the nucleotide binding domain are propagated either through the circumferential paths of subunit-subunit interfaces, or through the substrates such as DNA (portal protein and helicase) and γ-subunit (F_1 motor). The cooperativity among subunits is accomplished by the change of affinities to substrates or ligands coupled to the chemical reactions. Thus, the conserved motifs around the β-sheet, or the *engine* parts, play the same mechanical role in all of these biomolecular motors.

REFERENCES

[1] J. P. Abrahams, A. G. W. Leslie, R. Lutter, and J. E. Walker, "Structure at 2.8-Angstrom Resolution of F1-Atpase from Bovine Heart-Mitochondria," *Nature*, vol. 370, pp. 621-628, 1994.
[2] H. Y. Wang and G. Oster, "Energy transduction in the F-1 motor of ATP synthase," *Nature*, vol. 396, pp. 279-282, 1998.
[3] S. S. Patel and K. M. Picha, "Structure and function of hexameric helicases," *Annual Review of Biochemistry*, vol. 69, pp. 651-697, 2000.
[4] M. R. Singleton, M. R. Sawaya, T. Ellenberger, and D. B. Wigley, "Crystal structure of T7 gene 4 ring helicase indicates a mechanism for sequential hydrolysis of nucleotides," *Cell*, vol. 101, pp. 589-600, 2000.
[5] Y. J. Jeong, D. E. Kim, and S. S. Patel, "Kinetic pathway of dTTP hydrolysis by hexameric T7 helicase-printase in the absence of DNA," *Journal of Biological Chemistry*, vol. 277, pp. 43778-43784, 2002.
[6] H. Noji, R. Yasuda, M. Yoshida, and K. Kinosita, "Direct observation of the rotation of F-1-ATPase," *Nature*, vol. 386, pp. 299-302, 1997.
[7] S. Sun, D. Chandler, A. R. Dinner, and G. Oster, "Elastic energy storage in beta-sheets with application to F-1-ATPase," *European Biophysics Journal with Biophysics Letters*, vol. 32, pp. 676-683, 2003.
[8] G. Oster and H. Y. Wang, "Reverse engineering a protein: the mechanochemistry of ATP synthase," *Biochimica Et Biophysica Acta-Bioenergetics*, vol. 1458, pp. 482-510, 2000.
[9] D. E. Smith, S. J. Tans, S. B. Smith, S. Grimes, D. L. Anderson, and C. Bustamante, "The bacteriophage phi 29 portal motor can package DNA against a large internal force," *Nature*, vol. 413, pp. 748-752, 2001.

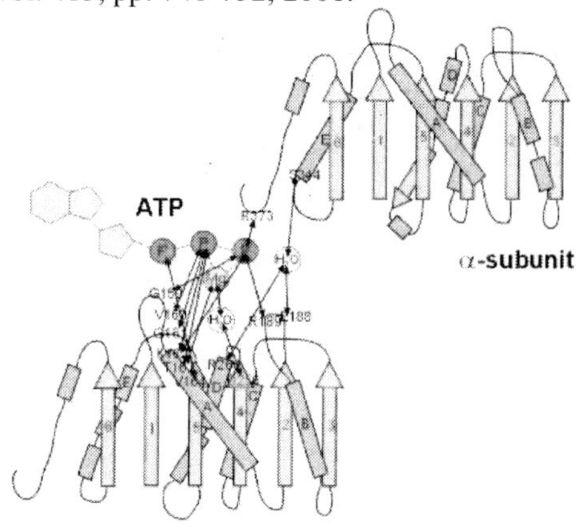

Fig. 1 Plot of the secondary structures of two adjacent F_1 subunits and an Mg-ATP molecule. The hydrogen bonds between subunits and Mg-ATP are shown.

A B

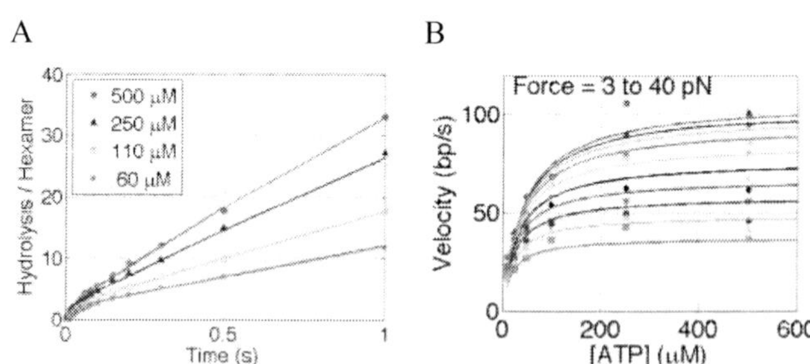

Fig. 2 The agreements of simulations and experiments for two motor proteins. (A) The transient kinetics of T7 helicase hydrolysis. Each curve corresponds to a different initial [dTTP]. (B) The steady-state translocation velocities of φ29 portal protein under various [ATP]. Each curve corresponds to a different external pulling force.

NANO2004-46006

THERMAL CONDUCTIVITY REDUCTION IN NANOSTRUCTURED SEMICONDUCTOR USING BROAD-BAND-PHONON SCATTERING

Woochul Kim
Department of Mechanical Engineering
University of California – Berkeley
Berkeley, CA 94720

Pramod Reddy
Applied Science and Technology
University of California – Berkeley
Berkeley, CA 94720

Arun Majumdar
Department of Mechanical Engineering
University of California – Berkeley
Berkeley, CA 94720

Joshua Zide
Department of Materials
University of California – Santa Barbara
Santa Barbara, CA 93106

Arthur Gossard
Department of Materials
University of California – Santa Barbara
Santa Barbara, CA 93106

Gehong Zeng
Department of Electrical and Computer Engineering
University of California – Santa Barbara
Santa Barbara, CA 93106

John Bowers
Department of Electrical and Computer Engineering
University of California – Santa Barbara
Santa Barbara, CA 93106

Ali Shakouri
Department of Electrical Engineering
University of Calfornia – Santa Cruz
Santa Cruz, CA 95064

EXTANDED ABSTRACT

Low thermal conductivity is essential for efficient operation of thermoelectric/thermionic power generation devices. There have been several attempts to design materials with low thermal conductivity without sacrificing electrical transport. These approaches utilized different mechanisms of phonon scattering, such as acoustic impedance mismatch of the adjacent layers in superlattices or defect scattering of phonons etc [1, 2]. However, each of these approaches scatter phonons only in a particular region of the phonon spectrum. In this paper we present experimental results of the thermal conductivity of epitaxially grown superlattices engineered to take advantage of the various scattering mechanisms to scatter phonons over the entire phonon spectrum.

Epitaxially grown superlattices of ErAs/InGaAs were used in our experiments. Figure 1 shows the cross sectional view of the ErAs/InGaAs superlattice structure. The black dots in the picture correspond to ErAs islands and the dark gray layer corresponds to InGaAs. This superlattice effectively scatters phonons in various regions of the phonon spectrum: a) The high frequency phonons in the superlattice are scattered by the InGaAs alloy structure present in the superlattice b) the low frequency phonons are preferentially blocked at the ErAs/InGaAs interface due to the acoustic impedance mismatch at the interfaces c) the intermediate part of the phonon spectrum is blocked by placing ErAs islands with dimensions comparable to the phonon wavelength.

Thermal conductivity of ErAs/InGaAs superlattice was measured by using the 3ω technique [3]. Figure 2 shows the experimental results of the thermal conductivity measurements of the ErAs/InGaAs superlattice. The thermal conductivity of Si-doped InGaAs is shown as a comparison. For the thermal conductivity of ErAs/InGaAs, a period thickness has been fixed as 40nm to see the effects of monolayer (ML) thickness over the thermal conductivity. Incorporation of ErAs scattering islands results in a significant reduction in the thermal conductivity possibly due to suppression of low and intermediate frequency phonons in addition to alloy scattering. However, the dependency of thermal conductivity on ML thickness of ErAs is less pronounced.

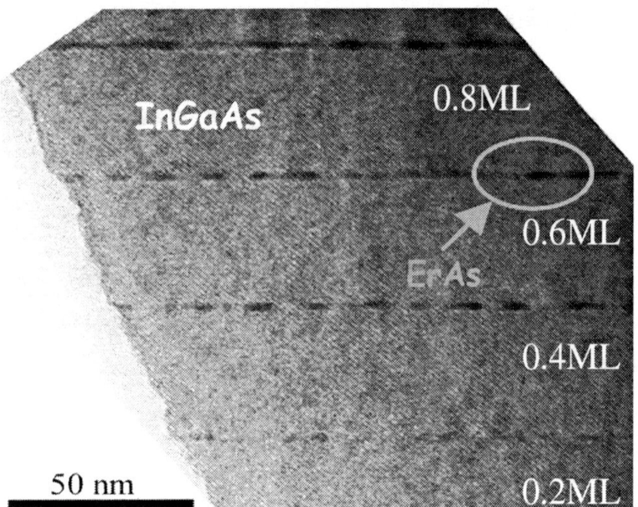

Figure 1: The cross sectional view of ErAs/InGaAs superlattice. High frequency phonons can be blocked by InGaAs alloy structure. The interface between ErAs and InGaAs can scatter low frequency phonons. The intermediate part of the phonon spectrum can be suppressed by the ErAs island itself. ML stands for monolayer. (Image from Elisabeth Muller, Paul Scherrer Institut Woerenlingen and Villigen, Switzerland)

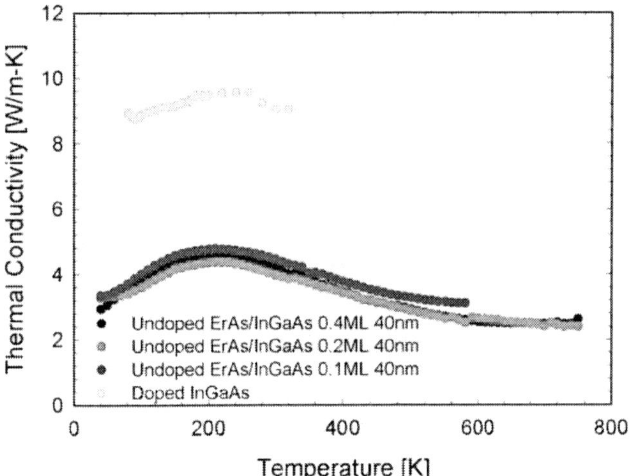

Figure 2: Thermal conductivity of ErAs/InGaAs supterlattice. The thermal conductivity of Si-doped InGaAs is shown as a comparison. Incorporation of ErAs scattering island has a significant effect on the thermal conductivity reduction. Thermal conductivity has less dependency over ML thickness of ErAs.

The effect of period thickness over the thermal conductivity is shown in Figure 3. Thermal conductivity measurement was done on 0.1 ML with different period thickness. One has 10nm period thickness and the other has 40nm. Although the total thickness of the superlattice is not same with each other, (1.2 μm and 1 6 μm respectively), the effect of period thickness over thermal conductivity do exist. ErAs/InGaAs superlattice with low period has less thermal conductivity. However the thermal conductivity increases again as in the case of 0.05ML with 5nm period thickness even though its low period thickness. It maybe the sizes of the ErAs islands are so a small compared with phonon wavelength, so that the scattering is now in the Rayleigh regime, where the phonon transport is already blocked by alloy scattering. Therefore, the low and intermediate wavelength phonons do not see any obstacles for their transport through superlattice. The thermal conductivity suppression is most evident between around 150 K and 400K. At high temperature above 600K, Umklapp scattering dominates over other scattering process.

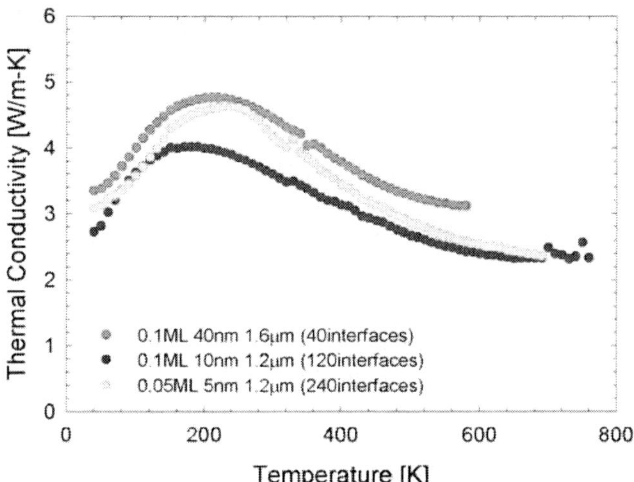

Figure 3: Low period thickness ErAs/InGaAs superlattice has a less thermal conductivity as in the case of 0.1 ML with different period thickness. However the thermal conductivity increases again in the case of 0.05ML 5nm period thickness.

Systematic approach combined with precise modeling will be pursued to fully understand the phonon transport in ErAs/InGaAs superlattice.

ACKNOWLEDGMENTS

This work is supported by the Office of Naval Research (ONR) - Multidisciplinary University Research Initiative (MURI) Grant.

REFERENCES

[1]   Venkatasubramanian, R., 2000, "Lattice thermal conductivity reduction and phonon localizationlike behavior in superlattice structures," *Physical Review B*, Vol. 61, pp. 3091-3097.

[2] S. T. Huxtable, A. Abramson, A. Majumdar, C.-L. Tien, G. Zeng, C. LaBounty, A. Shakouri, J. E. Bowers, and E. Croke, 2002, "Thermal conductivity of Si/SiGe and SiGe/SiGe superlattices," *Applied Physics Letters*, Vol. 80, pp. 1737 – 1739.

[3] Cahill, D. G., 1990, "Thermal conductivity measurement from 30 to 750 K: the 3ω method," *Review of scientific instrument*, Vol. 61, pp. 802-808.

NANO2004-46019

SELF-ASSEMBLY LIMITS IN STRUCTURAL PROTEINS

Bradley E. Layton, Drexel University, Department of Mechanical Engineering and Mechanics

ABSTRACT

A mechanics-based model is presented which predicts a "neutral annulus" for quasi-crystalline self-assembling nanostructure such as collagen fibrils, wherein transcript length, torsional and axial stiffness along with primary and quaternary protein structure limit the size to which these structures may aggregate. In the present treatment, a neutral annulus is predicted at 0.625 of the fibril radius, wherein portions of the fibril interior to the neutral annulus are in compression, balanced by portions exterior to the neutral annulus which are in tension.

INTRODUCTION

Collagen, perhaps the most abundant protein on earth in terms of mass, comprises approximately 7% of human body weight. Collagen has twenty distinct genes all of which are related by a tri glycine repeat G-X-Y where X is frequently proline (Boot-Handford and Tuckwell, 2003). The collagen gene is found in organisms as simple as *c elegans* (Adams *et al.*, 2000) indicating that it has been present in the unigenome long prior to vertebrate evolution. The predominantly occurring fibrillar collagen is Type I, a heterotrimer consisting of two identical and one distinct alpha helix. The common triglycine repeat theme, unique to collagen and responsible for its ability to form a triple helix was certainly a fortuitous event for multicellular life, in that it allowed for spatial organization of specialized cells. Its ability to self-assemble or aggregate is an attribute, rather than a detriment as is the case with other genes such as those involved in Huntington's where tri-peptide repeats within DNA lead to fouling aggregation (Chen *et al.*, 2002). The repetitive sequence of the fibrillar collagens, however has the fortuitous consequence of forming polymers which can span spatial dimensions several times the size of a typical cell and more than likely provided early eukaryotic cells a means for establishing the spatial organization required for multicellular life.

What this paper attempts to address is a baseline for predicting the radial aggregation limit for self-assembling polymers such as collagen. The energy-minimization method used herein is a first step in predicting the radial dimension to which a self-assembling polymeric quasicrystal grows. This work will not only aid in understanding the nanostructure healthy and diseased tissue, but will also provide a strategy for predicting tissue remodeling rates and tissue repair rates. The energy minimization method described herein also provides a relatively straightforward predictive tool for the engineering of self-assembling nanomaterials. The unique nature of collagen with its fixed transcript length of ~1,100 amino acids, resulting in a length of approximately 300nm is responsible for the well-known axial periodicity of 62-69nm. Several other biochemical hypotheses have been presented as predictors of the aggregation limit. It has been found, for example that both the N and C termini are responsible for stages of assembly but must be cleaved on order for mature fibrils to form (Hulmes, 2002). In the current work, these preliminary biochemical steps are modeled as necessary but not sufficient predictors of the fibril radial aggregation limit. At the quaternary scale, collagen's alpha helices form ~1.4nm diameter triple helices. These three alpha helices, sometimes genetically identical (e.g. Type III) and sometimes combinations of two or three genetic transcripts are bound by hydroxylysine crosslinks to form a right-handed triple helix. These self-assembling triple helices form a supraquaternary structure known as a fibril which is again right handed reaching diameters of 10-500nm and indeterminate lengths.

As the fibrils aggregate or self-assemble, what becomes clear is that in order to maintain the axial periodicity of 62-69nm while simultaneously maintaining the supraquaternary helical packing angle, as the fibril grows, triple helices at the periphery must strain in tension to accommodate the axial periodicity if the supraquaternary helical packing angle is to be maintained. As a preliminary test of this hypothesis (Figure 1) shows results of a metastudy, showing that diameter is inversely proportional to D-band spacing. An immediate conclusion from this is that as a fibril is growing, newly added and bound outer triple helices compress axially those interior to themselves until a stable or steady-state is reached. Presumably at this point the binding energy of additional unbound or partially bound triple helices is not great enough to overcome the additional tensile energy required to reach the requisite binding sites and the fibril stops growing.

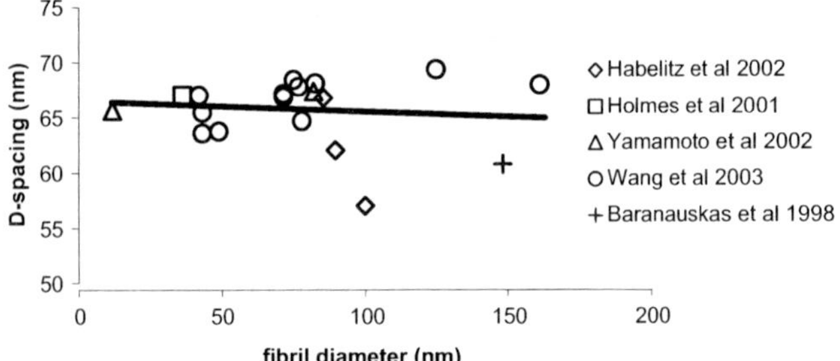

Figure. 1. Results of a metastudy show that collagen fibril D-spacing is inversely proportional to fibril diameter (Baranauskas *et al.*, 1998; Holmes *et al.*, 2001; Habelitz *et al.*, 2002; Yamamoto *et al.*, 2002; Wang *et al.*, 2003).

The hypotheses which this model will attempt to explore are that (1) triple helices at the center of the fibril are in compression and triple helices closer to the periphery are in tension, (2) that the fixed transcript length of collagen, its discrete binding sites, and its torsional and axial stiffness are sufficient predictors for the collagen fibril diameter aggregation limit.

NOMENCLATURE

symbol	meaning	typical value
d	triple helix diameter	1.4nm
D	axial periodicity of fibril	57-69nm
i	radial index	
j	circumferential index	
L	triple helix length	300nm
m	number of annuli in compression	
n	number of annuli in tension	
N	number of annuli in fibril ($n+m$)	~5-200
r	radial dimension	
s	circumferential distance between triple helix ends	
α	ratio between axial distance between triple helix ends at neutral annulus and axial periodicity (L_m/D)	4.4
γ	molecular axis tilt	5-30°
θ	circumferential dimension	

METHODS

A three-dimensional framework for predicting the collagen fibril radial aggregation limit is constructed using the four variables: (1) triple helix diameter, (2) D-spacing, (3) supraquaternary angle, and (4) triple helical modulus. For the model being presented, torsional stiffness and knowledge of binding sites are neglected and are reserved for future work.

Assume that a collagen fibril has radial symmetry and that a given triple helix lies roughly within a given radial plane (or annulus). For visualization purposes, first imagine making a slice from the perimeter of the fibril to its center running the entire length of the fibril and opening it up so that each annulus is planar (Fig 2).

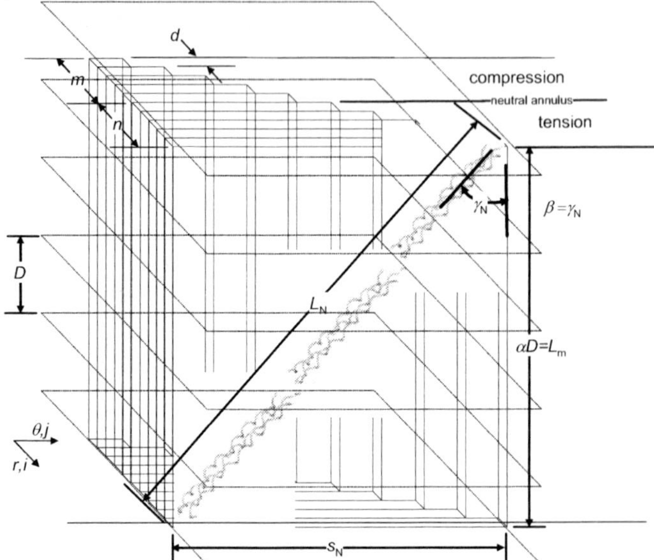

Figure 2. 3D rendering of unrolled fibril used to model collagen fibrils as a nested set of annularly arranged triple helices. As triple helices of fixed transcript length are added to the periphery, their propensity to bind is limited by the distance between binding sites. For a given supraquaternary angle, b, if D-spacing is to be maintained, aggregating triple helices must stretch. This has the effect of compressing inner triple helices, reducing D-spacing, and limiting the diameter to which a fibril may grow.

Each triple helix lies a distance $r_i = di$ from the center of the fibril where d is the triple helix diameter (~1.4nm) and i is its radial index. The number of annuli is N. The number of annuli in compression is m and the number of annuli in tension is n such that $m+n=N$, with the mth annulus representing the neutral annuls. We adopt the familiar notation of D representing the axial periodicity and let L_i represent the triple helical length in the ith annulus where L_m is the relaxed length of the triple helix. A relationship may be established where the axial compressional energy in the fibril is equal and opposite to the axial tensile energy.

$$U_C = U_T . \qquad (1)$$

This may be rewritten as

$$E_{3H} \sum_{i=1}^{m} \sum_{j=1}^{f(i)} \frac{\varepsilon_{i,j}^2}{2} = E_{3H} \sum_{i=m+1}^{N} \sum_{j=1}^{f(i)} \frac{\varepsilon_{i,j}^2}{2}, \qquad (2)$$

where E_{3H} is the triple helix elastic modulus, $\varepsilon_{i,j}$ is the strain of the triple helix in the ith annulus and j^{th} circumferential position in the fibril and $f(i)$ is a function of the annular position of the triple helix approximated by

$$f(i) = 2\pi r_i = 2\pi i (r_{3H}). \qquad (3)$$

Since the fibril has radial symmetry, allow each triple helix within a given annulus to have the same strain and let

$$\varepsilon_{i,j} = \frac{L_i - L_m}{L_m} = \frac{L_i}{L_m} - 1, \qquad (4)$$

where L_i is the length of a triple helix in the ith annulus and L_m is the length of a relaxed triple helix. The length of every other triple helix may be expressed as:

$$L_i = \frac{\alpha D}{\cos \gamma_i} = \frac{L_m}{\cos \gamma_i}, \qquad (5)$$

where γ_i is the angle that the triple helix in the ith annulus makes with the fibril axis and α is the ratio between the triple helix length and the D-spacing.

$$\gamma_i = \tan^{-1} \frac{s_i}{\alpha D} \qquad (6)$$

where s_i is the circumferential distance between the two ends of a triple helix in the ith annulus

$$s_{i,j} = \frac{r_i}{R} \alpha D \tan \beta \qquad (7)$$

where all j triple helices of length $s_{i,j}$ in annulus i, are assumed to be equal due to radial symmetry. R is the fibril radius. Experimentally the supramolecular angle, β, at the surface may be measured, and thus, at the fibril perimeter, $\beta = \gamma_N$. Substituting into Eq. 2,

$$\sum_{i=1}^{m} \sum_{j=1}^{f(i)} \left(\frac{1}{\cos\left(\tan^{-1}\left(\frac{r_i - r_m}{R} \tan \beta \right) \right)} - 1 \right)^2 =$$

$$\cdot \sum_{i=m+1}^{N} \sum_{j=1}^{f(i)} \left(\frac{1}{\cos\left(\tan^{-1}\left(\frac{r_i - r_m}{R} \tan \beta \right) \right)} - 1 \right)^2 \qquad (8)$$

where the radial index, i runs from 1 to m, the location of the neutral annulus, on the LHS and from $m+1$ to N, the total number of annular rings in the fibril on the RHS. Thus the LHS represents the total compressive energy in the fibril and the RHS represents the total tensile energy. Note that the energy in the mth annulus actually contributes no energy.

As a preliminary verification that this method does predict a neutral annulus, let $N = \{1,2,3\ldots20\}$, $\beta = 5°$, and allow the neutral annulus to vary from the center to the perimeter in increments of $0.1R$.

RESULTS

The above mechanics-based model, predicts a mechanism for limiting collagen fibril diameter in tissues. Plotting energy in each annulus as a function of neutral annulus location results in the traces shown in Figure. 3.

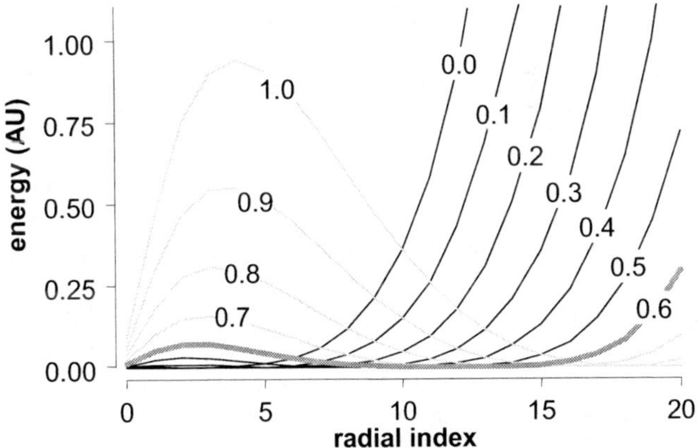

Figure 3. Total energy in model collagen fibrils as a function of the location of the neutral annuls. Concave down portions of the energy curves on the left side of the graph represent portions of the fibril in compression, while concave up portions of the energy curves on the right side of the graph represent portions of the fibril in tension. The curve with the least total energy in this figure is the one shown in red, where the neutral annulus is located at 0.6R, or 60% of the total fibril radius. Fibrils with the neutral annulus at the center (0.0) have all tensile annuli, while fibrils with the neutral annulus at the perimeter (1.0) have all compressive annuli.

Yellow represents annuli exterior to the neutral axis (tension); black represents annuli interior to the neutral axis (compression); and red is that predicted by the model at approximately $0.6R$. Further numerical results (not shown) predict the neutral annulus to be approximately $0.625R$. Integrating each of the curves of Figure 3, and plotting total energy as a function of location of neutral annulus (Figure 4), the minimum energy position is seen to occur between $0.6R$ and $0.7R$. The curve is biased toward lower energy values at the periphery, a possible indication that if second order effects such as hydration or internal entropic disorganization are present, then the neutral annulus may exist further out than the location predicted by this model.

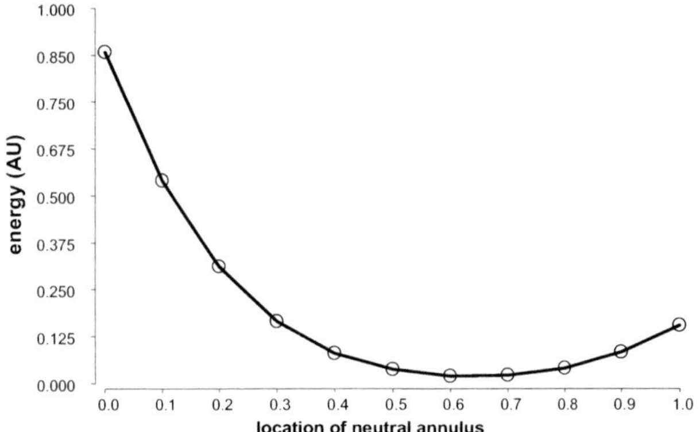

Figure 4. Total energy in model collagen fibrils depends on the location of the neutral annulus. Curves from Fig. 3 are integrated and plotted to reveal that a neutral annulus near 0.6 results in a fibril with minimal energy, i.e, where compressive and tensile energies are equal.

DISCUSSION

This result matches recent empirical work (Gutsmann *et al.*, 2003) demonstrating that collagen fibril behavior is more tubelike than rod-like, in that the outermost portion of collagen fibrils appear to behave like a semi-collapsible tube, while the innermost portion appears to behave more like a fluid. This is consistent with the current model in that it predicts two distinct regions of the fibril. This work has further implications in the area of self-assembly of molecular structures at the nanoscale and may represent a naturally occurring mechanism for limiting the diameter to which collagen fibrils may grow. Future work which will include modeling binding energies, axial stiffnesses and torsional stiffnesses of individual collagen triple helices will provide a scale-dependent prediction of the actual aggregation limit. The current model, simply predicts the existence of a neutral annulus based on energy balance and geometrical considerations.

Potential implications of this work include the modeling of nanoscale growth and repair rates of tissue for the purpose of understanding aging, growth, tissue strength, tissue remodelability, and tissue engineering at the nanoscale. For example if one were to attempt to design a tissue from its constituent proteins, with collagen serving as the structural matrix, and wished to simplify the transcriptional and translational machinery by eliminating exons from the collagen genes altogether, the primary structure of the engineered protein would be of paramount significance since this would impact the quaternary structure and thus the supraquaternary structure. In other words, any defects, indeed, single point mutations in engineered collagen genes must be evaluated for their propensity to accelerate, retard or preclude self-assembly.

ACKNOWLEDGMENTS
This work was funded by Drexel University.

REFERENCES
Adams, M. D., S. E. Celniker,. G. M. Rubin, J. C. Venter et al. (2000). "The genome sequence of Drosophila melanogaster." Science **287**(5461): 2185-95.

Baranauskas, V., B. C. Vidal, N. A. Parizotto (1998). "Observation of geometric structure of collagen molecules by atomic force microscopy." Appl Biochem Biotechnol **69**(2): 91-7.

Boot-Handford, R. P., D. S. Tuckwell (2003). "Fibrillar collagen: the key to vertebrate evolution? A tale of molecular incest." Bioessays **25**(2): 142-51.

Chen, S., F. A. Ferrone, R. Wetzel (2002). "Huntington's disease age-of-onset linked to polyglutamine aggregation nucleation." Proc Natl Acad Sci U S A **99**(18): 11884-9.

Gutsmann, T., G. E. Fantner, M. Venturoni, A. Ekani-Nkodo, J. B. Thompson, J. H. Kindt, D. E. Morse, D. K. Fygenson, P. K. Hansma (2003). "Evidence that collagen fibrils in tendons are inhomogeneously structured in a tubelike manner." Biophys J **84**(4): 2593-8.

Habelitz, S., M. Balooch, S. J. Marshall, G. Balooch, G. W. Marshall, Jr. (2002). "In situ atomic force microscopy of partially demineralized human dentin collagen fibrils." J Struct Biol **138**(3): 227-36.

Holmes, D. F., H. K. Graham, J. A. Trotter, K. E. Kadler (2001). "STEM/TEM studies of collagen fibril assembly." Micron **32**(3): 273-85.

Hulmes, D. J. (2002). "Building collagen molecules, fibrils, and suprafibrillar structures." Journal of Structural Biology **137**(1-2): 2-10.

Wang, H., B. E. Layton, A. M. Sastry (2003). "Nerve collagens from diabetic and nondiabetic Sprague-Dawley and biobreeding rats: an atomic force microscopy study." Diabetes Metab Res Rev **19**(4): 288-98.

Yamamoto, S., J. Hitomi, S. Sawaguchi, H. Abe, M. Shigeno, T. Ushiki (2002). "Observation of human corneal and scleral collagen fibrils by atomic force microscopy." Jpn J Ophthalmol **46**(5): 496-501.

NANO2004-46036

TRANSPORT PHENOMENA IN NANOFLUIDIC CHANNELS

Hirofumi Daiguji
Institute of Environmental Studies
The University of Tokyo
7-3-1 Hongo, Bunkyo-ku, Tokyo
113-0033, Japan

Peidong Yang
Department of Chemistry
University of California, Berkeley
Berkeley, CA 94720-1460

Andrew Szeri
Department of Mechanical Engineering
University of California, Berkeley
Berkeley, CA 94720-1740

Arun Majumdar
Department of Mechanical Engineering
University of California, Berkeley
Berkeley, CA 94720-1740

INTRODUCTION

Ion transport in nanoscale channels has recently received increasing attention. Much of that has resulted from experiments that report modulation of ion transport through the protein ion channel, α-hemolysin, due to passage of single biomolecules of DNA or proteins [1]. This has prompted research towards fabricating synthetic nanopores out of inorganic materials and studying biomolecular transport through them [2]. Recently, the synthesis of arrays of silica nanotubes with internal diameters in the range of 5-100 nm and with lengths 1-20 µm was reported [3]. These tubes could potentially allow new ways of detecting and manipulating single biomolecules and new types of devices to control ion transport. Theoretical modeling of ionic distribution and transport in silica nanotubes, 30 nm in diameter and 5 µm long, suggest that when the diameter is smaller than the Debye length, a unipolar solution of counterions is created within the nanotube and the coions are electrostatically repelled [4]. We proposed two different types of devices to use this unipolar nature of solution, i.e. 'transistor' and 'battery'. When the electric potential bias is applied at two ends of a nanotube, ionic current is generated. By locally modifying the surface charge density through a gate electrode, the concentration of counterions can be depleted under the gate and the ionic current can be significantly suppressed. This could form the basis of a unipolar ionic field-effect transistor. By applying the pressure bias instead of electric potential bias, the fluid flow is generated. Because only the counterions are located inside the channel, the streaming current and streaming potential are generated. This could form the basis of an electro-chemo-mechanical battery. In the present study, transport phenomena in nanofluidic channels were investigated and the performance characteristics were evaluated using continuum dynamics.

THEORETICAL MODELING

In the analysis of ionic current across a channel, the Poisson-Nernst-Planck (PNP) equations and the Navier-Stokes (NS) equations were employed.

$$\nabla^2 \phi = -\frac{1}{\varepsilon_0 \varepsilon} \sum_a z_a e n_a , \tag{1}$$

$$\nabla \cdot \left(n_a \boldsymbol{u} + \boldsymbol{J}_a \right) = 0 , \tag{2}$$

$$\nabla \cdot \boldsymbol{u} = 0 , \tag{3}$$

$$\boldsymbol{u} \cdot \nabla \boldsymbol{u} = \frac{1}{\rho} \left\{ -\nabla p + \mu \nabla^2 \boldsymbol{u} - \left(\sum_a z_a e n_a \right) \nabla \phi \right\} . \tag{4}$$

where ε_0 is the permittivity of vacuum, ε is the dielectric constant of medium, n_a is the concentration of ions of species a, $z_a e$ is their charge, ρ is the fluid density and μ is the viscosity. In eq 2, $\boldsymbol{J}_a$ denotes the particle flux due to concentration gradient and electric potential gradient, which is given by

$$\boldsymbol{J}_a = -D_a \left(\nabla n_a + \frac{z_a e n_a}{kT} \nabla \phi \right) \tag{5}$$

where D_a is the diffusivity of ion species a. The boundary conditions at channel walls are as follows.

$$\nabla_\perp \phi = -\frac{\sigma}{\varepsilon_0 \varepsilon} , \tag{6}$$

$$J_{a\perp} = 0 , \tag{7}$$

$$\nabla_\perp p = \mu \nabla_\perp^2 u - \left(\sum_a z_a e n_a \right) \nabla_\perp \phi , \tag{8}$$

$$\boldsymbol{u} = \boldsymbol{0} . \tag{9}$$

where $\perp$ denotes the wall-normal component and σ is the surface charge density. In this study, KCl aqueous solutions were used. The dielectric constant of KCl aqueous solution, ε, is 80, the diffusivities of K^+ and Cl^-, D_{K+} and D_{Cl^-}, are 1.96×10^{-9} and 2.03×10^{-9} m^2/s, respectively. Figure 1 shows the 2D domain for the calculation, where the total length of the channel L_x is 5 μm and the height, L_y, is 30 nm, the reservoirs of 1 μm×1 μm are located at two ends of a channel. Temperature is 300 K, and the bulk concentration of KCl aqueous solution varies from 10^{-5} to 10^{-2} M.

For a nanofluidic transistor, because the fluid flow does not occur, only the PNP equations were solved. A gate electrode is located at the center of the channel and the length L_{gx} is 2 μm. The surface charge densities inside the channel, $\sigma=-10^{-3}$ and -2×10^{-3} C/m^2, whereas the surface charge density at the gate, σ_g is different from that of other parts due to the presence of the gate electrode. The electric potentials and the concentrations at the boundaries are given by constant values.

For the boundary conditions of a nanofluidic battery, when the electromotive forces, that is, the potentials at zero current are calculated, the gradients of electric potential are given by zero, and when the current-potential characteristics are calculated, the electric potentials are given by constant values. The gradient of velocity is given by zero and the pressure and concentrations adopt the bulk values.

RESULTS AND DISCUSSION

Figure 2 shows the effect of gate electrode on the ionic currents of K$^+$ and Cl$^-$. The bulk KCl concentration is 10^{-4} M and the potential bias is 5V. For both cases $\sigma=-10^{-3}$ and -2×10^{-3} C/m^2, as $\sigma_g\rightarrow0$, the ionic current is blocked almost completely. However, when $\sigma_g\sim\sigma$, the magnitude of the ionic current approaches the value at σ rapidly in comparison with those of σ_g in a whole channel wall, suggesting that the effect of σ_g on the ionic current is nonlinear. The ionic current of Cl$^-$ decreases monotonically with decreasing the surface charge density at the gate. As σ on the remaining surface increases from -10^{-3} to -2×10^{-3} C/m^2, the unipolar nature of charge transport appears more clearly. As $\sigma_g\rightarrow0$, the K$^+$ ions are depleted from under the gate such that the concentration drops to that of Cl$^-$. As a result most of the potential drop occurs at the gate. Since J_{K+} is constant along the channel, its value anywhere along the channel is determined by the depleted concentration under the gate.

Figure 3 shows the current-potential (I-$\Delta\phi$) curves of a battery. The pressure bias is 0.5 MPa, and the surface charge density is $\sigma=-10^{-3}$ C/m^2 on the channel surface. When the bulk concentration is large, the channel does not become a unipolar solution of counterions, and the output power is small. On the other hand, when the bulk concentration is so much small, the mass diffusion becomes a rate-controlling step and the potentials drop rapidly in the high current density region. There is an optimum concentration to generate the maximum efficiency of energy conversion. The Debye length of the

Figure 1. Computational grid (316×45) inside a silica nanotube with height of 30 nm, length of 5 μm with 1 μm × 1 μm reservoirs on either side. The gate is centered along the length of the nanotube with a length, L_{gx}=2 μm.

optimum concentrated solution is in the order of the height of a channel.

REFERENCES

[1] Kasianowicz, J. J., Brandin, E., Branton, D. and Deamer, D. W., 1996, "Characterization of Individual Polynucleotide Molecules Using a Membrane Channel," PNAS, 93, pp. 13770-13773.

[2] Li, J., Stein, D., McMullan, C., Branton, D., Aziz, M. J. and Golovchenko, J., 2001, "Ion-Beam Sculpting at Nanometre Length Scales," Nature, 412, pp. 166-169.

[3] Fan, R., Wu, Y., Li, D., Yue, M., Majumdar, A. and Yang, P., 2003, "Fabrication of silica nanotube arrays from vertical silicon nanowire templates," J. Am. Chem. Soc., 125, pp. 5254-5255.

[4] Daiguji, H., Yang, P. and Majumdar, A., 2004, "Ion Transport in Nanofluidic Channels," Nano lett., 4, pp. 137-142.

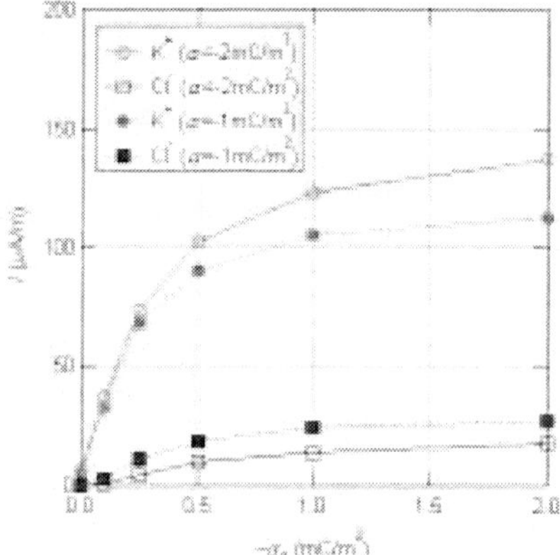

Figure 2. The I-σ_g characteristics. The charge densities on the channel surface except for the gate region are -2×10^{-3} (solid lines) and -10^{-3} C/m^2 (dashed lines). The bulk concentration of KCl aqueous solution is 10^{-4} M and the potential bias is 5V.

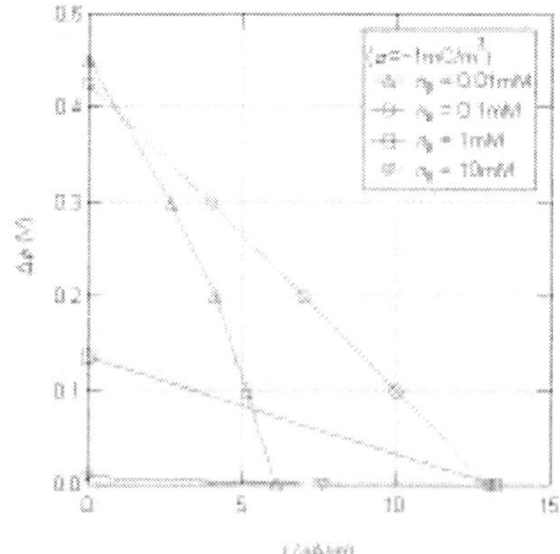

Figure 3. Current-potential (I-$\Delta\phi$) curves. The pressure bias is 0.5 MPa, and the surface charge density is $\sigma=-10^{-3}$ C/m^2 on the channel surface. The bulk concentrations of KCl aqueous solutions are assumed to be 10^{-5}, 10^{-4}, 10^{-3}, and 10^{-2} M.

NANO2004-46039

FOUNTAIN-PEN-BASED LASER MICROSTRUCTURING
WITH GOLD NANOINKS

Tae Y. Choi and Dimos Poulikakos

Institute of Energy Technology
Swiss Federal Institute of Technology Zurich
CH-8092, Switzerland

Costas P. Grigoropoulos

Department of Mechanical Engineering
University of California, Berkeley
Berkeley, CA 94720

ABSTRACT

Employing the fountain pen principle, a micropipette is used to write an Au nanoparticle ink on glass substrates. A continuous-wave laser (488-515 nm) is subsequently used as a controlled, localized energy source to evaporate the carrier liquid (toluene) in the ink and sinter the nanoparticles together thus fabricating continuous gold stripes 5 micrometers in width and a few hundred nanometers in height. The scanning speed, the laser intensity, and the degree of defocusing are identified as important parameters to the successful manufacturing of the gold microstructures. The electrical resistivity of the stripes, within the parametric domain of the present work, is measured to be the order of 10^{-6} ohm-m.

INTRODUCTION

A novel method of producing microstructures by sintering at temperatures markedly lower than the melting temperature of bulk materials involved, has been recently successfully demonstrated for gold nanoparticles [1-3]. The melting temperature of free-standing (i.e. not in a solvent medium) is expected to be in the range of 300-400 °C due to thermodynamic size effect [4]. Thus, nanoparticles can be effectively melted by utilizing localized laser heating, minimizing the damage on temperature sensitive substrates. The reduction of the total volume in conjunction with the vivid motion of the particles due to evaporation of the solvent (toluene) and strong absorption of the laser beam leads to a narrow gold stripe.

The present work focuses on the writing and laser annealing in tandem of a gold nanoparticle ink by utilizing the 'fountain pen' principle for the writing process. The nanoparticle solution was prepared with 30 wt % and 2 vol % of gold in toluene. The average diameter of the particles is in the range of 2-4 nm. The laser annealing is performed with the help of an inverted microscope setup.

RESULTS AND DISCUSSIONS

The writing device is a simple pipette made to function as a 'fountain pen' (Fig. 1a). The pipette has been pulled out of a capillary tube (1.5 mm of outer diameter and 0.86 mm of inner diameter) by using a DMZ Universal Puller (Zeitz Instrumente). The optimal size of the opening was found to be about 21 μm, negating the use of pressurization to push the liquid from the pipette. The constant feed of gold suspension is possible due to the greater wetting force at the substrate than at the pipette end. One of the important factors for uniform and reproducible writing is the surface flatness of the pipette opening.

For curing and in-situ monitoring of the process (both writing and curing), an in-house inverted microscope with 50X objective lens (NA=0.5) was built. A continuous-wave Ar ion laser (Spectra Physics, Inc.) with multiple lines of 488-515 nm has been chosen for curing the gold lines since these lines are close to the absorption peak of the nanoparticle suspension at around 520 nm Assuming that the size of the laser beam is about the same as the objective lens aperture, the laser beam waist is estimated at around 1 μm.

A series of images illustrating the wetting of the substrate by the gold solution are shown in Fig. 1b. Fringe patterns signify that the gold solution at the pipette end starts contacting the substrate. The last snapshot depicts onset of wetting. Thereby, the writing process can be initiated by translating the computer controlled XY microstages. A line written by the pipette is shown in Fig. 1c by maintaining the distance between the pipette and the substrate (soda-lime glass) at about 2 μm. To achieve this, the sample tilting is assured within a couple of micrometers (~depth of focus of the objective) at the interval of 10 mm by using a tilting stage and the microscope. A piezo stack actuator is used to measure the exact tilting amount represented by applied voltage and compensated during the writing process by a LabVIEW controlled program, enabling the constant gap between the pipette and the substrate. The size is 13 μm larger than the pipette opening of 21 μm due to the wetting on the substrate. The cross section of the written gold solution, as shown in Fig. 1d, is obtained by utilizing the shear force microscopy technique with scanning speed of 20 μm/sec (0.1 Hz scanning frequency). The contact angle is measured at 7.5 degrees from Fig. 1d. The precisely controlled laser (size of 1 μm in $1/e^2$ diameter) is irradiated in the middle of the printed line to cure the gold nanoparticles. Washing away the uncured solutions at the sides, the width of the cured gold stripe of 5 μm has been achieved with laser power and curing speed at 8.8 mW and 25 μm/sec, respectively (Fig 2). As shown in the AFM scanned cross-section, the line thickness is revealed measured to be 400 nm. The profile is smooth with roughness of about 20 nanometers (peak to peak) except for the edges where thermocapillarity driven by Marangoni flow migrates the particles of the uncured region toward the outer edge of the cured line where they accumulate and agglomerate.

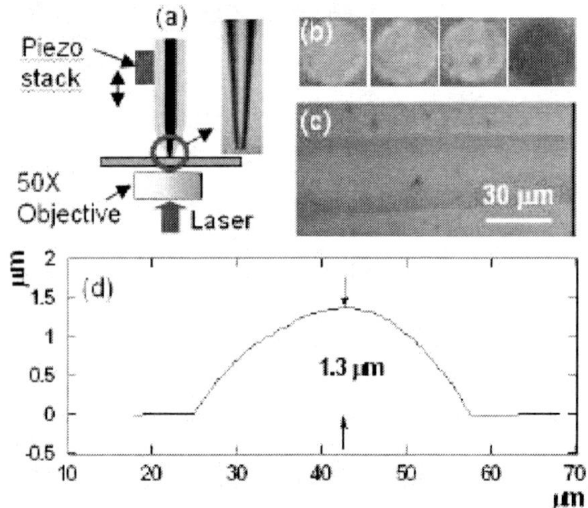

Fig.1 (a) Experimental setup for writing and laser curing, (b) Light microscopy images of wetting of the substrate, (c) Au nanoparticles ink written on the substrate, and (d) cross sectional view of the line in (c) by scanning force microscopy.

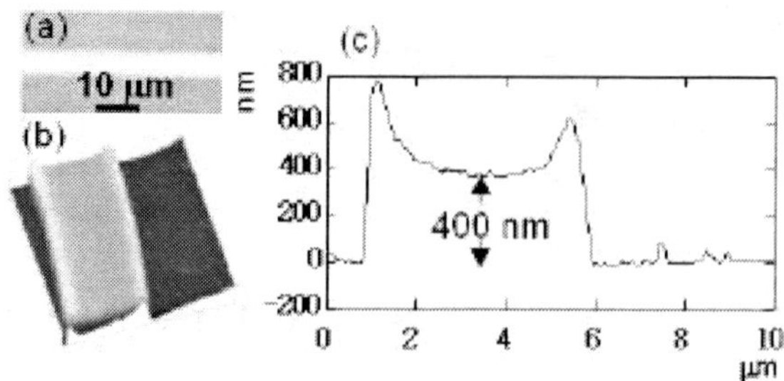

Fig. 2 (a) Light microscopy images of a cured line by laser, (b) AFM scanned image of (a), and (c) its cross section. Laser power and scanning speed are at 8.8 mW and 25 µm/sec, respectively.

References:

1. N. Bieri, J. Chung, S. E. Haferl, D. Poulikakos, and C. P. Grigoropoulos, Appl. Phys. Lett. **82**, 3529, 2003.
2. J. Chung, S. Ko, N. Bieri, C. P. Grigoropoulos, and D. Poulikakos, Appl. Phys. Lett. **84**, 801, 2004.
3. J. Chung, N. Bieri, S. Ko, C. P. Grigoropoulos, and D. Poulikakos, Superlattices and Microstructures, accepted for publication, 2004.
4. P. Buffat and J. P. Borel, Phys. Rev. A **13**, 2287, 1975.

NANO2004-46041

SIMULATION OF THERMIONIC EMISSION FROM QUANTUM WIRES

Yang Liu and T.S. Fisher[#]
School of Mechanical Engineering and Birck Nanotechnology Center
Purdue University
West Lafayette, IN 47907
[#]tsfisher@purdue.edu

EXTENDED ABSTRACT

In the late 19th century, Edison observed electrical current flowing between hot and cold electrodes [1]. Since this discovery of thermionic emission, research has occurred with varying intensity in order to harness the simplicity and utility of the thermionic effect in power generation devices. Hatsopoulos and Gyftopoulos [2,3] provide details of the development of thermionic theory and practice. In general, thermionic power generation has not found widespread use, despite many inherent advantages over alternative power generation methods, because of material limitations that have precluded an attractive combination of power generation efficiency and capacity. This paper presents semiclassical and quantum models for the thermionic behavior of a newly developed class of materials, quantum wires, that may offer some promise in alleviating historic materials limitations of thermionic devices.

A general expression for the emitted current as a function of field, temperature and work function is established for a quantum wire based on a semi-classical approach, and the results differ from those of 3-D bulk materials. Figure 1 shows a schematic of the problem geometry in which a wire is attached to a bulk metallic material at one end. The other end is a bulk electrode separated by a vacuum gap. The wire's axis is parallel to the direction of thermionic emission. Simulation of thermionic emission from a quantum wire is achieved from a quantum transport point of view with the non-equilibrium Green's function (NEGF) method, which includes relevant mesoscopic physics and has been widely applied to transport problems in nanostructures [4].

The value of the NEGF formalism lies in providing a physically sound model that accurately represents open boundary conditions, where interfacial transport effects between bulk and confined conductors are important. In the NEGF formalism, the quantum wire and the vacuum part are described by the Hamiltonian matrix. The effect of coupling between the quantum wire and the semi-infinite contact is described though a self-energy matrix, whose anti-Hermitian part, the broadening

matrix, is related to the finite lifetime of electronic states, reflecting the fact that an electron introduced into a state does not remain there forever, but leaks away into the contact. The advantage of introducing the self-energy matrix lies in the fact that a very large matrix is needed to describe the whole "quantum wire and semi-infinite bulk" system and the matrix dimension for the bulk is so large that makes any real computation impossible.

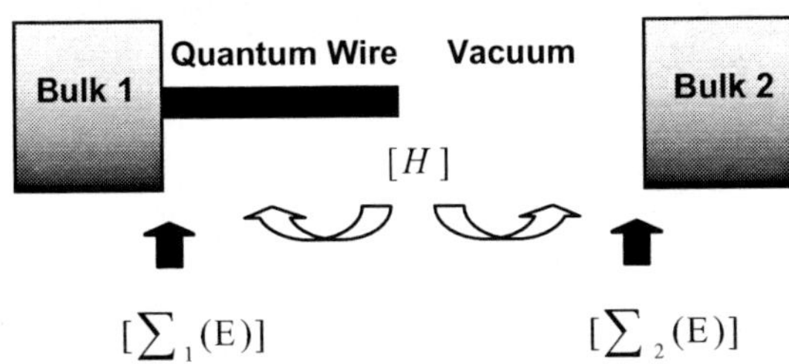

FIGURE 1. Schematic of thermionic emission from a quantum wire. The Hamiltonian is identified by $[H]$. Self-energy matrices $[\Sigma_1(E)]$, $[\Sigma_2(E)]$ account for coupling to the two bulk contacts.

The introduction of the self-energy matrix is not only mathematically convenient but also conceptually important. When the quantum wire is brought into contact with the semi-infinite bulk, the local density of states (LDOS) of the wire is broadened into a Lorentzian form due to the wave functions spilled over from the bulk. To describe the broadening effect and boundary condition accurately, the self-energy matrix is included in the wire's Hamiltonian matrix, which has the same dimension as that of the wire. The self-energy matrix consists of the coupling matrix between the quantum wire and the bulk and the surface Green's function matrix of the contact. For a simple 1-D problem, only one element of the self-energy matrix is nonzero, accounting for the contact point exactly. The two semi-infinite contacts give two different Fermi functions when they exist at different temperatures and act as an electromotive force in a circuit that induces the current flow. The thermionic current is obtained by integrating over all energy levels the product of the difference between two Fermi functions and the transmission function, which represents the probability of electron transmission through or over a barrier.

Theoretical and simulated results match very well in the semi-classical regime [5]. At the same time, both results indicate a higher current density and thus higher energy conversion capacity than that of a bulk material with the same work function. Thus, thermionic emission from quantum-confined materials might provide a technical path toward improving the efficiency and capacity of direct energy conversion devices and systems.

Thermionic emission current densities for bulk emitters obtained from the quantum transport theory (NEGF) and semi-classical approach are compared in Fig. 2. The predicted current densities for bulk emitters are similar for all temperatures. Further, simulated current densities using the NEGF approach indicate that thermionic current density decreases exponentially with increasing work function (not shown), as the semi-classical approach predicts.

The fact that the emission current from a bulk metal predicted by NEGF is always smaller than that given by the semi-classical approach results from the difference in transmission functions. The semi-classical approach models transmission as a step function in which all electrons with energies that exceed the surface potential barrier (or work function) are emitted. However, the transmission function given by NEGF is less than unity for energies slightly above the work function because the inclusion of wave-like (quantum) transport causes a non-zero probability of reflection for particles that exceed the energy barrier.

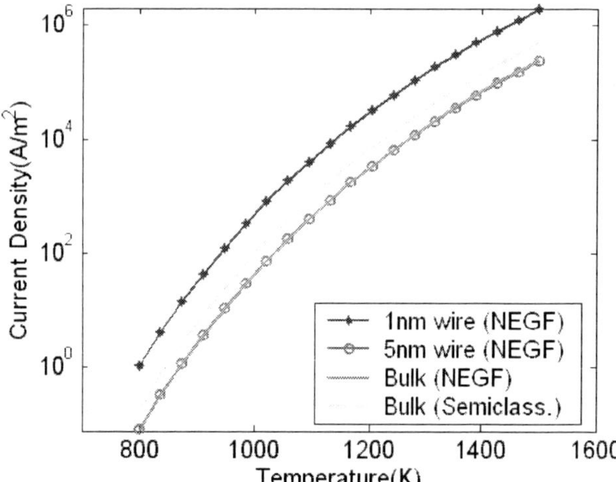

FIGURE 2. Thermionic current density as a function of emitter temperature for NEGF and semi-classical approach. Work function $\Phi = 2.0\,\mathrm{eV}$ for all temperatures. The temperature of the other bulk electrode is fixed at 300K.

The emission current densities from quantum wires are also compared to that from bulk emitters in Fig. 2. The results indicate a substantial increase in current density for the smallest wire (1nm diameter) due to quantum confinement, which enhances the occupation of high-energy electrons in the emission direction. For the larger wire (5nm diameter), the current density matches that of a bulk emitter. Thus, this size effect indicates that larger thermionic emission current density can be achieved for wires with diameters ~1nm, and arrays of such quantum wires could be applied in future devices and systems for improving the efficiency and capacity of direct energy conversion.

Relevant issues concerning the feasibility of using quantum-wire thermionic emission devices include selective surface termination for lowering work functions (and characterization thereof), achieving high-density vertical nanowire arrays, and enhanced understanding of interfacial transport between bulk and confined electron conductors, and these are the subject of ongoing research.

ACKNOWLEDGMENTS

We gratefully acknowledge funding from the NASA-Purdue Institute for Nanoelectronics and Computing, and the NSF Nanoscale Interdisciplinary Research Team programs.

REFERENCES

1. W.H. Preece, "On Peculiar Behavior of Glow-Lamps when Raised to High Incandescence," Proceedings of the Royal Society (London), Serial A, Vol. 38, p. 219, 1885.
2. G.N. Hatsopoulos and E.P. Gyftopoulos, Thermionic Energy Conversion, Volume I, Massachusetts Institute of Technology, Cambridge, MA, 1973.
3. G.N. Hatsopoulos and E.P. Gyftopoulos, Thermionic Energy Conversion, Volume II, MIT Press, Cambridge, MA, 1979.
4. S. Datta, "Nanoscale device modeling: the Green's function method," Superlattices and Microstructures, 28, 253-278, 2000.
5. T.S. Fisher, "Theory of Thermionic Emission from a Quantum Wire," Sixth ISHMT/ASME Heat Transfer Conference, Kalpakkam, India, 6 pp., January 2004.

66

NANO2004-46059

SIMULATION OF THERMAL TRANSPORT IN NANOWIRE COMPOSITES FOR MACRO-ELECTRONICS APPLICATIONS

Satish Kumar
School of Mechanical Engineering
Purdue University
585 Purdue Mall
W. Lafayette IN 47907-1288
kumar21@purdue.edu

Jayathi Y. Murthy
School of Mechanical Engineering
Purdue University
585 Purdue Mall
W. Lafayette IN 47907-1288
jmurthy@purdue.edu

M.A. Alam
School of Electrical and Computer
Engineering
Purdue University
465 Northwestern Avenue
W. Lafayette IN 47907-2035
alam@purdue.edu

ABSTRACT

Thermal transport in thin film transistors (TFTs) composed of nanowires embedded in plastic substrates is considered. Random ensembles of intersecting and contacting wires embedded in a substrate are analyzed using Fourier theory. Heat generation due to self-heating is included. A finite volume scheme is used to obtain the temperature solutions in the wires and substrate. Temperature profiles in the ensemble are investigated as a function of wire number density, wire-contact Biot number as well as the Biot number for heat transfer to the substrate.

INTRODUCTION

Thin film transistors (TFTs) built on plastic substrates are increasingly being explored for use in flexible, wearable and disposable electronics. Polycrystalline Si TFTs are difficult to implement on plastics because of the high processing temperature required [1]. Amorphous Si semiconductors can be processed at lower temperatures, but do not have sufficient carrier mobility to be viable. On the other hand, single nanotube and nanowire field effect transistors (FETs) are being explored which have demonstrated very high mobility. One possibility is to use nanowires (NWs) or tubes made of high-mobility materials and embed them in plastic. Recently, a low-temperature process was used to fabricate Si NW TFTs in plastic [1] and favorable device characteristics were reported.

A potential bottleneck in developing transistors based on plastic-embedded nanowires and nanotubes is heat dissipation. Heat generated due to electron-phonon scattering must be dissipated either through the low-conducting plastic or along the length of the wire. If the wires are strongly oriented, there is little wire-to-wire contact. On the other hand, if large areas are to be covered, it is likely that perfect alignment cannot be achieved, and significant wire-to-wire contact would exist. In such a situation, heat transport through the matrix of nanowires

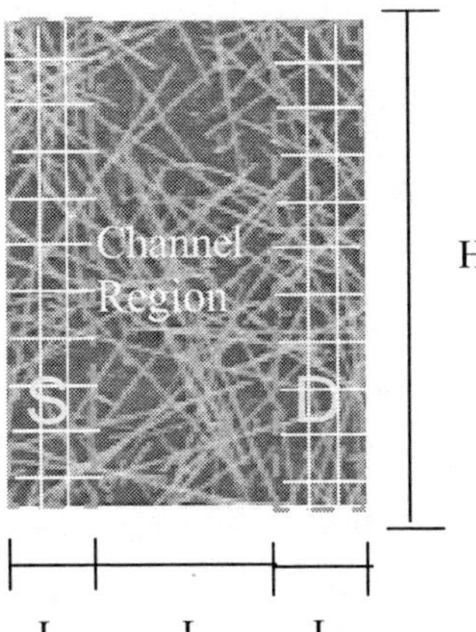

Figure 1: Computational domain showing nanowires, source (S), drain (D) and channel region

could potentially provide additional heat transfer pathways and ameliorate the self-heating problem.

The objective of the present work is to quantify the extent of self-heating in nanowire matrices embedded in a plastic substrate. The geometry is shown in Fig. 1. A rectangular domain contains dispersed nanowires embedded in plastic. The temperatures in the source and drain region are assumed held at T_C. Nanowires of length l and diameter d are randomly dispersed in the domain. Some wires span the source-to-drain distance L, while others may not, depending on orientation. Those that span the width L carry current and hence have a volumetric heat generation q'''.

GOVERNING EQUATIONS

Though microscale heat transfer effects such as phonon ballistic transport, boundary scattering and confinement are likely to be important, Fourier conduction is assumed in this analysis as a starting point. Using the dimensionless variables $\theta = (T - T_c)/(q''' l d / k_{wire})$ and non-dimensionalizing all lengths by the wire diameter d, the governing equations in the wires and substrate are written as:

$$\frac{d^2\theta_i}{ds^{*2}} + Bi_c\left(\theta_j - \theta_i\right) + Bi_s\left(\theta_s - \theta_i\right) + \frac{d}{l} = 0 \tag{1a}$$

$$\nabla^{*2}\theta_s + \sum_{i=1}^{N_{wires}} Bi_s\beta_v\left(\theta_i - \theta_s\right) = 0 \tag{1b}$$

Here, θ_i (s^*) is the temperature of the ith wire at axial location s^*, and $\theta_s(x^*,y^*,z^*)$ is substrate temperature. Thermal contact between wires i and j is characterized by the contact Biot number Bi_c. Heat exchange between each wire and the substrate is governed by substrate Biot number Bi_s and the contact parameter β_v. These are defined as:

$$Bi_c = \frac{h_c P_c d^2}{k_{wire} A}; \quad Bi_s = \frac{h_s P_s d^2}{k_{wire} A}; \quad \beta_v = \alpha_v\left(\frac{A}{P_s}\right) * \frac{k_{wire}}{k_{substrate}}$$

Here, h_c and h_s are the heat transfer coefficients for characterizing wire-to-wire and wire-to-substrate contact, P_c and P_s are the corresponding contact perimeters, k_{wire} is the thermal conductivity of the wire and A its cross-sectional area. The parameter β_v characterizes contact geometry and α_v is the contact area per unit volume of substrate. For low volume fractions, Bi_s may be computed assuming cylindrical wires in an infinite medium [2]. Bi_c is more difficult to characterize and requires detailed contact modeling. It is treated as a parameter in the present study. Additional dimensionless parameters governing the problem are the conductivity ratio $k_{wire}/k_{substrate}$, wire aspect ratio d/l, as well as the geometric ratios L_1/l, L/l and H/l.

NUMERICAL METHOD

The random ensemble of nanowires is generated in the following way. The source and drain regions in Fig. 1 are divided into finite control volumes. A probability p=0.1 of any control volume originating a nanowire, is chosen *a priori*. A random number is picked from a uniform distribution and compared with p. If it is less than p, a nanowire is originated from the control volume. The orientation of the wire is also chosen from a uniform random number generator. Since the wire length is fixed at *l*, all wires may not span the width of the channel region L.

The finite volume method [3] is used to solve the temperature field in the wires and the substrate. Each wire is divided into 1-D segments, and the finite volume scheme is applied. Possible contact with other wires is checked for each segment. Bi_c is zero for the segment if there is no contact. The substrate is divided into 3-D control volumes, and the finite volume scheme is used for discretization. Segments of the nanowires located in each substrate control volume are identified and their heat loss is ascribed to the control volume to ensure conservation. The discrete equations for all the nanowires and the substrate are solved sequentially and iteratively until convergence.

RESULTS

Dimensionless wire temperatures for the case d/L=0.01, L_1/l =1.0, L/l =0.5; H/l=2.0 and Bi_s=0 are computed for a range of Bi_c values in Fig. 2. A total of 90 wires are used in the domain. The computations show that nanowire contact does indeed provide a cooling mechanism for high Bi_c. A maximum temperature wire θ=9.75 is predicted for Bi_c=0 (no contact); in contrast, θ=4.28 is predicted for Bi_c =10. The heated region is more spread-out for higher Bi_c but the peak temperature is lower.

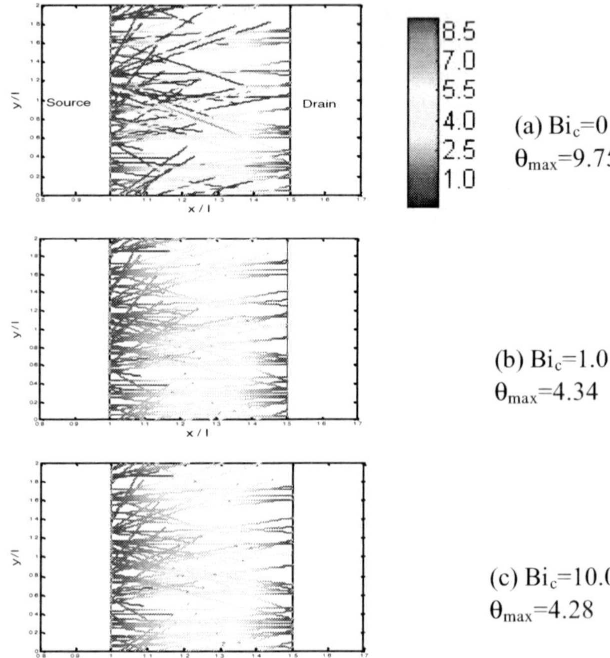

Figure 2: Temperature distribution along nanowires

CLOSURE

Heat transfer in ensembles of contacting nanowires embedded in a plastic substrate has been analyzed numerically using Fourier theory. A more complete exploration of the parameter space is underway, and extensions to incorporate sub-continuum effects as well as concurrent electro-thermal simulations are planned for the future.

REFERENCES

[1] Duan, X., Niu, C., Sahl, V., Chen, J., Parce, J.W., Empedocle, S., and Goldman, J.L., "High performance thin-film transistors using semiconductor nanowires and nanoribbons," *Nature*, Vol. 425, pp. 274—278, 2003.
[2] Morgan, V. T., "The overall convective heat transfer from smooth circular cylinders," Advances in Heat Transfer, Vol, 11, pp. 199 – 264, 1975.
[3] Patankar, S.V., Numerical Heat Transfer and Fluid Flow, Hemisphere, New York, 1980.

NANO2004-46066

SHRINKING OF GOLD-NANOINK PATTERN BY WATER INDUCED DEWETTING

Cédric P.R. Dockendorf, Tae-Youl Choi and Dimos Poulikakos
Laboratory of Thermodynamics in Emerging Technologies, Institute of Energy Technology
Swiss Federal Institute of Technology Zurich
CH-8092, Switzerland

ABSTRACT

Nanoinks show great potential for patterning of surfaces in small scales. Further reduction of pattern size, however, is limited in current inkjet and other related deposition technologies. We present a method to overcome this problem and reduce the size of deposited nanoparticle suspension films by water induced dewetting of the nanoink film.

NOMENCLATURE

σ interfacial tension
ϑ contact angle
L_i substrate-ink interface length
$L_{\hat{i}}$ substrate-surrounding medium interface length
C_i curve length of the ink-surrounding medium interface

INTRODUCTION

With many novel microelectronics applications, such as high density electronics, flexible electronics, disposable electronics for RFid tags (1) or biomedical tests, pushing towards further reduction of component size and manufacturing costs, a lot of effort is currently placed into the development of inkjet-inspired and other direct micro- and nanowriting techniques using nanoinks. A central issue in the above mentioned efforts, is the further reduction of the conductor size.

We present herein a method to reduce the pattern size of deposited gold nanoparticle suspensions through preferential evaporation of the toluene. Subsequent curing of the suspension produces electrical conductors. As has been shown by Bieri et al. (2) and Chung et al. (3), nanoparticle suspensions have great potential in micro- and nano-manufacturing. In accordance with the previous work described in reference (4), we use a borosilicate capillary tube filled with nanoinks, consisting of 4-8nm sized Au nanoparticles suspended in toluene as a fountain pen. The pipettes are pulled out of the capillary tubes using a DMZ Universal Puller (Zeitz Instrumente). The pipette opening is 20 µm, allowing for a constant ink flow due to a greater wetting force at the substrate than at the pipette opening. A flat pipette opening is essential for uniform nanoparticle suspension deposition. Prior to curing a water droplet is placed on top of the colloidal films. By heating the sample to a temperature slightly below the boiling point of water, toluene is evaporated

due to its higher vapor pressure. This evaporation of the toluene leads to substantial shrinking of the nanoinks film.

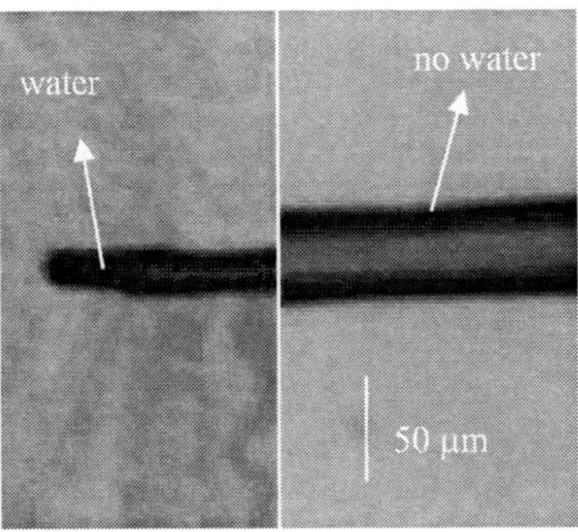

Figure 1: Film segment on the left: toluene was evaporated under a water film; Film segment on the right: toluene was evaporated in ambient air

Figure 1 shows a nanoink film which was heated to 95°C, one segment (the left half) was under a water film. The width of the segment which was heated under a water film was reduced from 50 µm to 20 µm, which corresponds to a volume decrease of 30.7 %. Measurements made by shear-force microscopy of the same segments confirm this result (figure 2). Comparing the profile of the shrunk pattern with the unchanged one (the right half of Fig. 1), one sees that an increase in contact angle took place, from 5.7° to 11.3°. As can be deduced from Young's equation at equilibrium:

$$\cos\theta = \frac{\sigma_{sv} - \sigma_{sl}}{\sigma_{lv}} \, ,$$

Equation 1: Young's Equation where σ: interfacial tension , θ: contact angle

the substitution of air by water as a surrounding medium of the liquid film and preferential evaporation of toluene result in a change of the contact angle.

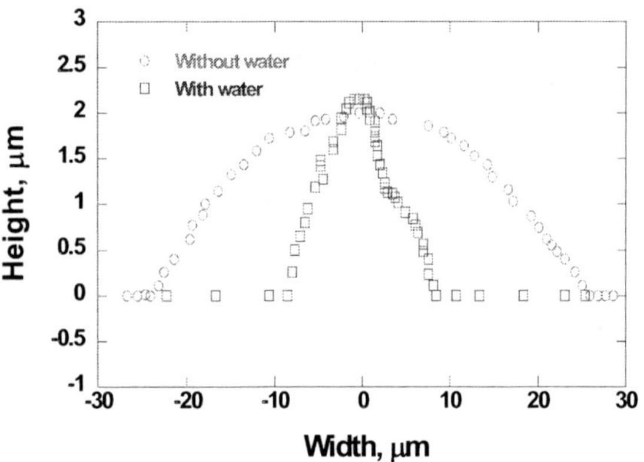

Figure 2:   Cross-sectional profiles of the lines by shear force microscopy

If the left hand side term, $\dfrac{\sigma_{sv} - \sigma_{sl}}{\sigma_{lv}}$, is less than -1, the three-phase contact line will be pushed towards the center of the film. After dewetting, the total surface/interface energy has to be less than the surface/interface energy before dewetting. The following relationship thus needs to be satisfied:

$$(C_a \cdot \sigma_{LV_a} - C_b \cdot \sigma_{LV_b}) + \left(L_a \cdot \sigma_{LS_a} - L_b \cdot \sigma_{LS_b}\right)$$
$$+ (L'_a \cdot \sigma_{SV} - L'_b \cdot \sigma_{SV}) \leq 0$$

Equation 2

where C is the curve length of the ink-water interface, L is the length of the ink-substrate interface and L' is the length of the substrate-surrounding medium interface. Assuming that σ_{SV} remains constant throughout the process and that σ_{LV} and σ_{LS} change, one can deduce that appropriate selection of the solvent and surrounding liquid can lead to an energetically driven dewetting process, where the shrinking drives towards the energy minimum.

The concentration of nanoparticles in the film changes as toluene evaporates from the solution, thus the viscosity of the film increases throughout the process. In accordance with Wang et al (5), one can assume that a higher viscosity inhibits dewetting. The concentration of particles in the suspension is also believed to have an impact on the final film size. The final equilibrium is thus expected to be a trade-off between a critical particle concentration and viscosity of the film.

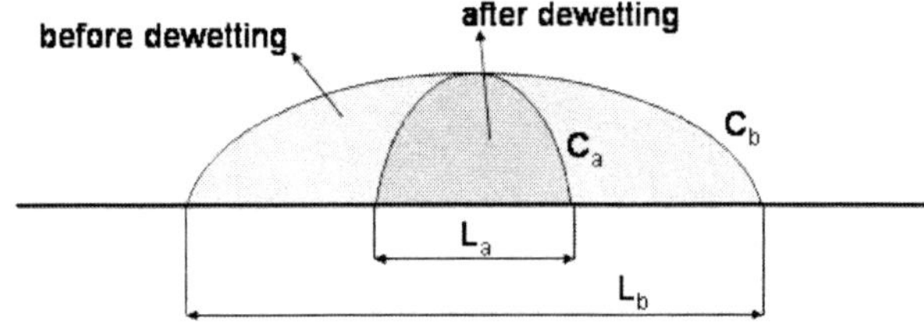

Figure 3: Schematic diagram of dewetting model

We have demonstrated that the width of a nanoparticle suspension film can be reduced by preferential evaporation of the solvent. This allows for further reduction in the size of electrical conductors manufactured by inkjet technologies and other micro-writing technologies.

REFERENCES

(1) D. Huang, F. Liao, S. Molesa, D. Redinger, V. Subramanian; Journal of the electrochemical society, vol 150, pp 412-417, 2003.
(2) N. Bieri, J. Chung, D. Poulikakos, C. P. Grigoropoulos ; Superlattices and Microstructures, 2004
(3) J. Chung, S. Ko, N.R. Bieri, C.P. Grigoropoulos, D. Poulikakos; Applied Physics Letters, Vol. 84(5), 2004.
(4) T.Y. Choi, D. Poulikakos, C. P. Grigoropoulos; Applied Physics Letters, 2004. To be published on June 28, 2004
(5) J.Z. Wang, Z.H. Zheng, H.W. Li, W.T.S. Huck and H. Sirringhaus; Nature Materials, vol 3, pp 171-176, 2004

NANO2004-46080

LOW-DIMENSIONAL PHONON HEAT CAPACITY OF TITANIUM DIOXIDE NANOTUBES

C. Dames and G. Chen
Department of Mechanical Engineering, Massachusetts Institute of Technology

B. Poudel, W. Wang, J. Huang, Z. Ren, Y. Sun, J. I. Oh, C. Opeil, S.J., and M. J. Naughton
Department of Physics, Boston College

Low-dimensional nanostructures such as nanotubes, nanowires, and quantum dots are promising building blocks for electronic, optical, sensing, and energy conversion applications. For effective device design it is important to understand how the basic thermal properties of nanostructures differ from those of bulk materials. For example, the measured thermal conductivity of silicon nanowires [1] can be understood with a 3-dimensional dispersion relation [2] for diameters down to about 40 nm, although at 22 nm diameter the experiment and modeling diverge sharply.

Because the phonon specific heat is due to the vibrational properties of the material, it is also a useful tool to study the basic properties of the lattice and nanostructure geometry. Previous studies of the specific heat in carbon nanotubes have shown various transitions from bulk to lower-dimensional behavior at temperatures T below ~50 K, with specific heat enhanced up to an order of magnitude greater than bulk graphite, although a coherent interpretation of the varied experimental results is still lacking [3, 4]. Previous studies of the specific heat in quantum dots have shown that quantum size effects on the allowed phonon wavevectors lead to smaller enhancements of up to 50% at low temperatures, followed by an exponential decay in the specific heat at even lower temperatures [5].

Titanium dioxide has important applications for Gratzel solar cells (anatase phase), gas sensing (rutile phase), photocatalysis (anatase), and supercapacitors (rutile). Nanotube geometries are of interest because of their high specific surface area, and greater understanding of their thermodynamic properties will facilitate practical applications. Here we present the first measurements of the specific heat of TiO_2 nanotubes (Fig. 1).

The TiO_2 nanotubes were synthesized by the standard chemical method from as-received bulk anatase powder (~100 nm particle size, 99.8% purity, Aldrich; and also ~44 μm particle size, 99.9% purity, Alfa Aesar) [6]. The resulting anatase nanotubes were typically 2.6 nm in wall thickness (4 layers), 10 nm in diameter, and ~600 nm long (Fig. 2). Different synthesis protocols led to large variations in crystallinity and impurity concentrations. Because the tubes are formed by the rolling up of a flat sheet, the interlayer coupling is expected to be non-covalent and therefore weak. The

nanotube samples were characterized using high-resolution transmission electron microscopy (HRTEM), selected-area electron diffraction (SAED), energy-dispersion x-ray analysis (EDX), scanning electron microscopy (SEM), and x-ray diffraction (XRD).

The specific heat was measured with a standard relaxation method in a liquid-helium cryostat and calorimeter (Oxford Instruments). Powders of as-synthesized nanotubes, or as-received bulk powder, were first cold pressed into pellets which were then cleaved into mm-scale samples of ~10 mg mass. The uncertainties in specific heat data are estimated as ±4% (±1 standard deviation) for the nanotubes and about ±10% for the bulk powders. The Cernox temperature sensor is believed to be accurate to better than ±1%.

Figure 1 shows selected results for bulk anatase and

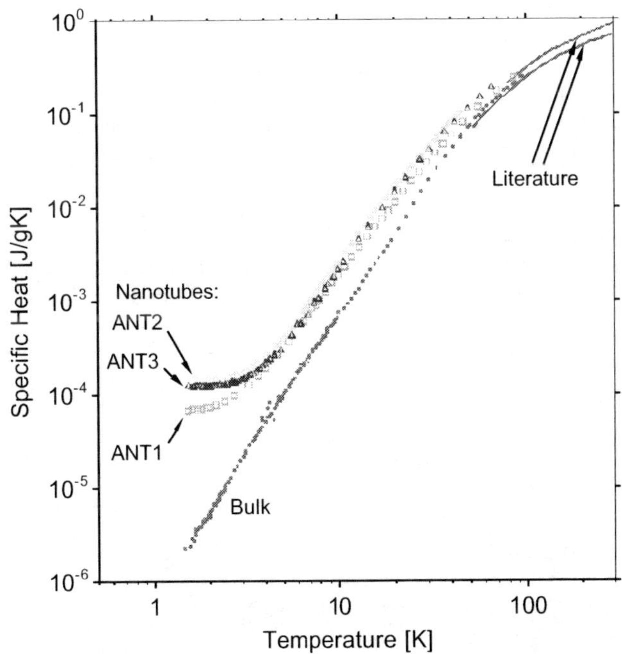

Figure 1. Measured specific heat of TiO₂ anatase nanotubes and bulk anatase, compared to bulk literature values [7, 8].

anatase nanotubes. The results for bulk rutile will be presented elsewhere. The bulk anatase sample is from the ~44 μm powder. The nanotube sample ANT1 has good crystallinity and good purity, ANT2 has good crystallinity but a moderate amount of Na impurities, and ANT3 has poor crystallinity (~50% amorphous) and poor purity (~10 at% Na impurities).

The bulk anatase specific heat data are in good agreement with literature values above 50 K [7, 8]. We are unaware of any literature values at lower temperatures. The data below ~45 K can be described to within ~10% by a T^3 law corresponding to a Debye temperature of 260 K, using the number density of primitive cells of $14.7*10^{27}$ m^{-3} (6 atom basis). This implies an averaged acoustic sound velocity of 3560 m/s.

The nanotube specific heat data always exceeds that of bulk anatase. From 95 K down to about 70 K, the excess is about 20 – 30%, with the pure anatase nanotube sample (ANT1) showing the least enhancement. From 50 K down to about 5 K, the nanotube curves shift to a ~$T^{2.6}$ dependence, clearly diverging from the ~T^3 behavior of bulk anatase. Below ~4 K the nanotube data shows a second transition to a nearly constant value, exceeding the bulk anatase values by factors of 25 to 50 at 1.5 K. Again, sample ANT1 shows the least enhancement, about half that of ANT2 and ANT3.

Quantitative modeling of these nanotubes is difficult because of: the complex crystal structure of anatase, with a polyatomic basis and unknown phonon dispersion relation even in bulk; the uncertain coupling between layers of individual tubes; the uncertain coupling between different tubes within a pellet; and in some samples the presence of impurities and amorphous phases. Nonetheless, the upper transition at ~50 K can be interpreted as a transition from 3-dimensional to 2-dimensional behavior, as the average phonon wavelength becomes comparable to the wall thickness. Assuming a constant sound velocity v_s, well below the Debye temperature the phonon analog of Wien's displacement law gives the peak thermal wavelength λ_T as

$$\lambda_T\, T \approx (50\ \text{nm K}) * (v_s / 5000\ \text{m/s}). \qquad (1)$$

Thus, for a typical sound velocity of 3700 m/s, at 50 K the characteristic wavelength is ~0.7 nm. Although this is relatively small compared to the typical wall thickness of 2.6 nm, if the layers are weakly coupled then it is the single-layer thickness of 0.9 nm, rather than the total wall thickness, that should be compared to the thermal wavelength to determine the dimensionality. However, this interpretation would predict T^2 (2-dimensional) behavior below 50 K, rather than the $T^{2.6}$ trend seen in Fig. 1.

The lower transition at ~3 K is even more difficult to interpret. The reasoning used above suggests a second transition when the thermal wavelength becomes comparable to the nanotube circumference. At 3 K the characteristic wavelength is ~12 nm, indeed becoming comparable to the circumference of ~27 nm. However, this interpretation would predict T^1 (1-dimensional) behavior below 3 K, rather than the nearly flat T^0 behavior seen in Fig. 1.

The nearly constant specific heat between 1.5 and 3 K could be explained by a large collection of modes with very low energy, as in a Debye or Einstein model with a characteristic temperature below ~1 K (~0.1 meV). This could be due to low-energy, long-wavelength vibrations between adjacent tubes, or to additional "tunneling" or "localized excitation" modes of atomic rearrangement. The latter mechanism is widely believed to be important for a T^1 behavior in amorphous glasses at low temperature [9], and recently has been suggested as a possible explanation for surprising $T^{0.34}$ and $T^{0.62}$ behavior below ~0.6 K reported for single wall carbon nanotubes [3].

In summary, these results show three different regimes of behavior for the anatase nanotubes compared to bulk anatase, delineated by transitions at ~50 K and ~3 K, and exceeding the bulk value by over 30 times at 1.5 K. The transition to nearly constant specific heat below ~3 K is particularly surprising and of fundamental scientific interest. Above ~70 K, the specific heat of anatase nanotubes converges to within 20% of bulk anatase. This important result affirms the guideline of Eq. 1: that even nanometer-sized systems may behave like bulk when the characteristic wavelengths are smaller than the sample dimensions. Because most of the interest in TiO_2 devices is at 300 K and above, it seems likely that the bulk specific heat would be an excellent approximation for the true nanotube specific heat in technological applications. To better understand the different contributions to the nanotube specific heat additional work would be required with a systematic variation of wall thickness and tube diameter, ideally on individual isolated nanotubes. For example, varying the wall thickness would aid the understanding of the upper transition at ~50 K, making it possible to estimate the strength of the interlayer coupling (covalent versus weakly bonded).

ACKNOWLEDGMENTS

This work was funded by NSF (NIRT-0210533), the US Army Research Development and Engineering Command, Natick Soldier Center (DAAD16-03-C-0052), and DOE (FG02-02ER45977).

REFERENCES

[1] Li, D, Wu, Y., Kim, P., Shi, L., Yang, P., and Majumdar, A., Appl. Phys. Lett. **83**, 2934 (2003).

[2] Dames, C. and Chen, G., J. Appl. Phys. **95**, 683 (2004).

[3] Lasjaunias, J. C., Biljakovic, K., Monceau, P., and Sauvajol, J. L., Nanotech. **14**, 998 (2003).

[4] Dresselhaus, M. S., and Eklund, P. C., Adv. in Phys. **49**, 705 (2000).

[5] Novotny, V., and Meincke, P. P. M., Phys. Rev. B **8**, 4186 (1973).

[6] Poudel, B., Wang, W., Dames, C., Huang, J., Kunwar, S., Wang, D., Banerjee, D., Chen, G., and Ren, Z. (submitted for publication).

[7] Touloukian, Y. S., ed., Thermophysical properties of matter, IFI/Plenum, New York, 1970.

[8] Wu, X.M., Wang, L., Tan, Z.C., Li, G.H., and Qu, S.S., J. Solid State Chem. **156**, 220 (2001).

[9] Phillips, W. A., ed., Amorphous Solids, in Topics in Current Physics **24**, Springer-Verlag, New York, 1981.

Figure 2. SEM (left) and TEM (right) images of typical TiO₂ nanotubes.

NANO2004-46083

MASS TRANSPORT PHENOMENA IN SUPERHYDROPHOBIC SURFACES

Wei Xu
University of California
Integrated Nanosystems Research Facility
Irvine CA 92697-2625
E-mail: xw233@hotmail.com
Mark Bachman
University of California
Integrated Nanosystems Research Facility
Irvine, CA 92697-2625
E-mail: mbackman@uci.edu

Hong Xue
California State Polytechnic University
Mechanical Engineering Department
3801 West Temple Ave., Pomona, CA 91768
E-mail: hxue@csupomona.edu
G.P. Li
University of California
Integrated Nanosystems Research Facility
Irvine, CA 92697-2625
E-mail: gpli@uci.edu

ABSTRACT

We present results of a droplet placed on a controlled super-hydrophobic surface cooled underneath by a thermal electrical cooler to demonstrate quick change in contact angles from the Cassie composite contact state to the Wenzel wetting contact state. The measured contact angles are compared with the theoretical predictions of Cassie's and Wenzel's equations and found to be consistent. The actual details of the transition phenomena are observed under a microscope through a specially designed one-dimensional micro-channel with concaved structures at the two sidewalls. It is found that the temperature gradient enhanced mass transfer can cause a rapid condensation in the air-filled cavities, which is believed to be the possible mechanism to trigger the energy state transition and explain instabilities of super-hydrophobic surfaces at the Cassie state. The phenomenon of mass transport into micro and nanocavities is important in understanding the nature of nano-structured super-hydrophobic surfaces.

1. INTRODUCTION

Recently, there has been great interest in generating nanopatterned surfaces to achieve a superhydrophobic condition at the surface [1, 2, 3]. The potential benefits for truly non-wetting surfaces include many industries such as aerospace and biotechnology, as well as emerging technologies such as microfluidics where surface interactions are undesirable. The most effective surface designs utilize high aspect ratio structures, which serve to. hold water drop on a composite surface, patchwork of solid and air. Such designs generate the Cassie condition, sometimes referred to as the "lotus leaf effect"—a condition in which very little water is contact with surface, resulting in very high contact angles approaching 180 degrees. Furthermore, water is seen to move freely with almost no interaction on such surfaces allowing it to sheet off the surface leaving no residue. Fabrication strategies have ranged from micromachining arrays of tiny posts to growing aligned nanotubes, which are chemically modified to be hydrophobic. However, most Cassie surfaces tend to be unstable and researchers have noted that after extended times, water drops will spontaneously change from a super-hydrophobic state to a less desirable Wenzel state, where drops of liquid are effectively fills up the micro structures on the rough surface.

From the surface energy consideration and the classical Young's equation on contact angle, the apparent contact angels θ_a^W and θ_a^C for the respective Wenzel and Cassie states can be derived as [4]

$$\cos\theta_a^W = r\cos\theta \qquad (1)$$

$$\cos\theta_a^C = \phi_s(\cos\theta + 1) - 1 \qquad (2)$$

where r is the ratio of the actual liquid-solid contact area to the projected area on the horizontal plane, θ is the equilibrium contact angle of the liquid drop on the flat surface, ϕ_s is the solid fraction of the liquid-solid contact.

The Wenzel state in Eq. (1) shows that the surface roughness can amplify the wetting. Since the ratio r is always larger than 1, for hydrophilic surfaces ($\theta < \pi/2$), the wetting gets expanded ($\theta_a^W < \theta$) and for hydrophobic surfaces ($\theta > \pi/2$), it gets shrunk ($\theta_a^C > \theta$). In fact, when the surface is hydrophobic, Patankar [5] showed that the surface energy of dry solid is lower than that of wet solid and thus it is expected that the equilibrium shape corresponding to the Cassie state will be encountered first. At a enhanced composite contact with small area fraction of liquid-solid contact ϕ_s, usually the apparent contact angle θ_a^C is larger then θ_a^W. Such an energy state is usually unstable, which leads to a undesired transition from the composite to wetting contact, although the details of how the energy barrier transit are not understood.

In this study, we demonstrate a nano-scale mass transport

condition for the change and the relation to the instability of super-hydrophobic state are investigated.

2. EXPERIMENTS

2.1 Preparation of Super-hydrophobic Surfaces

We manufactured two types of super-hydrophobic surfaces for this study. The first surface was a silicon surface with a two-dimensional array of high aspect ratio posts shown in Fig.1 (a), which was fabricated by standard photolithography and deep RIE process. The cylindrical posts are 3.7 μm in diameter and 70 μm in height (Fig.1 (b)) and the pitch between the posts is 11.4 μm. To treat the surface hydrophobic, it was soaked in sigma-cote (Sigma Chemical Co., St. Louis, MO), and put into vacuum for overnight. Finally the surface was taken out and totally blown dry. Drops of water placed on this surface displayed 165 degree contact angles in Cassie state and rolled easily across the surface. Secondly, we designed a special micro-channel for direct observation of the interface of water drop and the super-hydrophobic surface. A one-dimensional

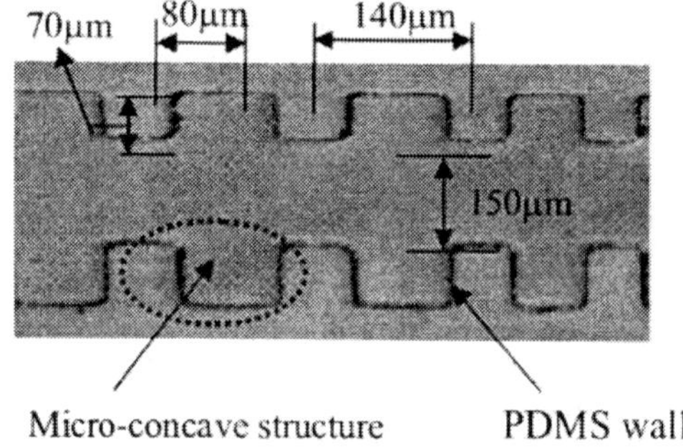

(a) SEM picture of micro-post surface and dimensions

(b) PDMS micro-channel with micro-concave-structure wall

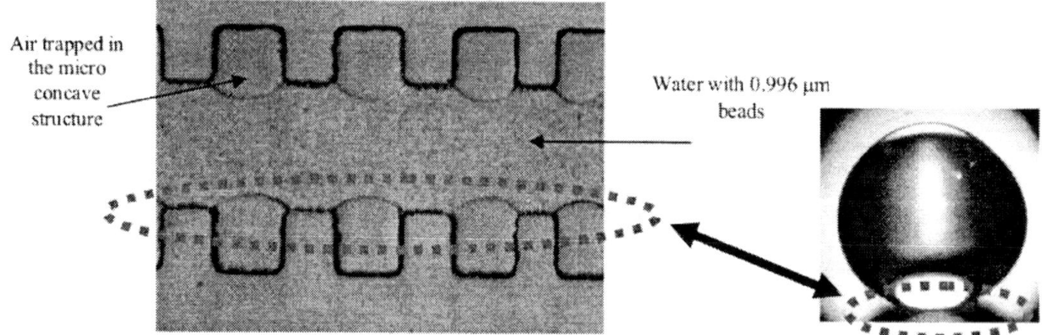

(c) PDMS microchannel with micro-concave-structure wall

Figure 1 Super-hydrophobic surfaces and micro channel

phenomenon that can cause a super-hydrophobic condition to change from the Cassie state to the Wenzel state. The trigging

equivalent in PDMS is generated for study of mass transport at liquid-solid and liquid-vapor composite contacts. This consists of a small channel 150 µm × 50 µm × 4.5 cm ($W{\times}D{\times}L$), with structured sidewalls containing concave cavities, 80 µm × 70 µm × 50 µm ($W{\times}H{\times}D$) shown in Fig.1 (c). The ratio of wall surface to concave surface is about 1 to 3. The channel was fabricated in PDMS by soft-lithography method casting from a micromachined mold.

When the micro-channel was filled with water, if the pressure is less than a critical value (about 4 kPa estimated from an experiment), the water placed in the channel would not fill the concave micro cavities. To make interface of water and air easily visible, beads of 0.996 µm in diameter (Duke Scientific Corp. Palo Alto, CA) were added into the water. Figure 1(c) demonstrated that the Cassie condition at the composite contact surface. This enabled us to produce super-hydrophobic micro channels and directly view the status of the air cavities under a microscope.

2.2 Experiments on Super-hydrophobic Surface

Silicon super-hydrophobic substrate was put on a thermal electric cooler (Advanced Thermoelectric Co., Nashua, New Hampshire, NH). Super-hydrophobic side was faced up. The cooler has a cooling area of 2 cm × 2 cm, which is the same size as the size of silicon substrate so that uniform cooling on the super-hydrophobic surface can be achieved. A water drop was formed by dispensing the liquid from above then the equilibrium shape corresponding to a composite contact was encountered. At the beginning, both water drop and super-hydrophobic surface were at room temperature, which was 26°C. Then the thermal electric cooler was turned on to cool down the super-hydrophobic surface. The temperature difference between surface and water drop was set as 4°C, 11°C and 16°C respectively. The images of water drop apparent contact angle were recorded with a digital camera mounted on a microscope though a reflection mirror. The digital camera was directly connected to a PC.

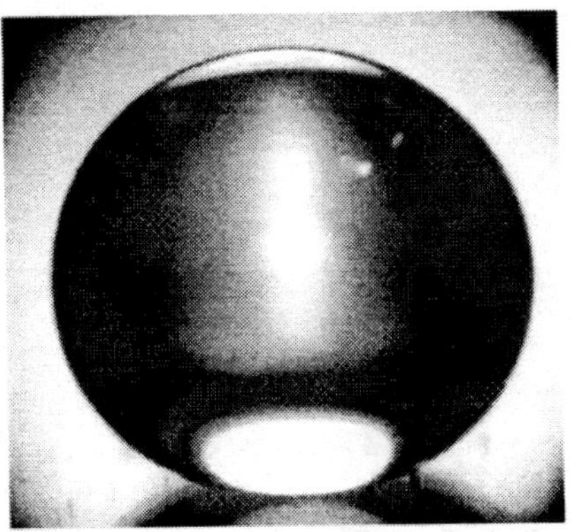

(a) Water droplet on the micro-post surface demonstrates a 165° contact angle; both temperature of droplet and surface are at 26°C

(b) Temperature of micro-post surface was reduced to 22°C; the contact angle of the droplet becomes 135° after 50s

(c) Temperature of micro-post surface was reduced to 15°C; the contact angle of the droplet becomes 130° after 30s

(d) Temperature of micro-post surface was reduced to 10°C; the contact angle of the droplet becomes 130° after 20s

Figure 2 Microscope images and contact angles of water droplet imposed to a cooled micro-post surface

2.3 Experiments on Micro-channel

We filled the micro-channel with 20 mM sodium phosphate (EMD Chemicals Inc., Gabbston, NJ) buffer containing beads 0.996 µm in diameter at an electrical field of 200V/cm applied at the two ends of micro-channel to heat the buffer by Joule heating. The images of the process at the composite contact in the micro-channel were recorded by a digital camera, which was connected to a PC

3. RESULTS AND DISCUSSIONS

3.1 Transition from the Cassie State to Wenzel State

When a small drop of 10 µl de-ionized water was first placed on the super-hydrophobic surface at room temperature of 26° C, the drop displayed a super-hydrophobic contact angle of 165° and moved freely on the surface as illustrated in Fig.2 (a). As the condition remained, there was no significant change observed in contact angle with a relatively long period of time (15 minutes). However, when the super-hydrophobic surface was cooled to three different temperatures, namely 22 °C,15°C, 10°C with a voltage-controlled thermal electric cooler, the contact angle dropped to 135°~130° within 50, 30 and 20 seconds respectively. The mobility of the water drop was also reduced considerably. This indicates that the initial composite contact in the Cassie state transited to a wetting contact of the Wenzel state with a reduction in contact angle and mobility. Figure 2 (b), (c) and (d) shows images of droplet and contact angles at three temperature differences. The cooler the surface is, the shorter the time period needs for the transition to take place.

3.2 Predictions of a Theoretical Model

The apparent contact angles according to Wenzel's and Cassie's equations (1) and (2) can be related to the geometric parameters of the surfaces roughness. The equations can be expressed as

$$\cos\theta_a^W = \left(1 + \frac{4A}{(d/h)}\right)\cos\theta$$

$$\cos\theta_a^C = A(\cos\theta + 1) - 1$$

where

$$A = \frac{\pi/4}{(1 + b/d)^2} \tag{3}$$

d is the diameter of the post, h is the height of the post and b is the pitch between the posts.

Figure 3 shows the apparent contact angles at both Cassie and Wenzel states for a hydrophobic rough surface. The apparent contact angle at the Cassie state is independent of d/h while the wetting contact angle at Wenzel state depends on d/h. Both curves are plotted against ratio of b/d, a parameter indicating the area fraction of the liquid-solid contact given in Eq. (3). Additional curves for $d/h = 0.5$ and 1.0 at Wenzel states are also plotted for reference purpose. In the case of micro-post

surface tested in the experiment, $b/d = 3.08$ and $d/h = 0.053$, indicated in an arrow in Fig. 3, the predicted apparent contact angle at the Cassie state is 164° ($\cos\theta_a^C = -0.961$), and the corresponding apparent contact angle at the Wenzel state state is 142° ($\cos\theta_a^W = -0.793$), which are reasonably consistent with experimental measurements. The transition from composite to wetting contact on the microscopic scale indicates an energy transport between two equilibrium positions, each holding its own energy barrier. It is also interesting to note from Fig. 3, there is a global minimum energy state [5] existing at the intersection of the curves of the Cassie and Wenzel states. Approaching to such a state requires either a very small of a/h ratio, which is limited by the capability of fabrication process or a small b/d ratio, which suggests that the diameter of the cylindrical posts needs to be at nanometer scale.

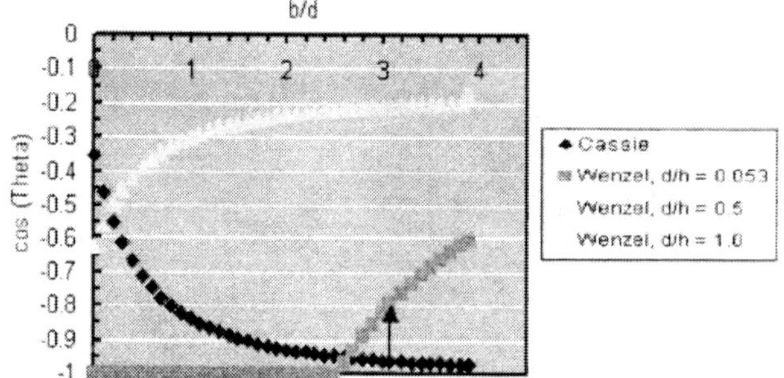

Figure 3 Predicted apparent contact angles at both Cassie and Wenzel states for a hydrophobic surface

3.3 Mass Transfer at Liquid-Vapor Interface

The specially designed micro-channel provides us a direct observation of actual details of the transition to understand the mechanism. As it is difficult to cool the channel uniformly, we filled the micro-channel with conductive buffer and ran a current through the buffer to generate Joule heating. As indicated in Fig.4, we observed almost immediate random formation of sub-micron water droplets in the air cavities with applied current.

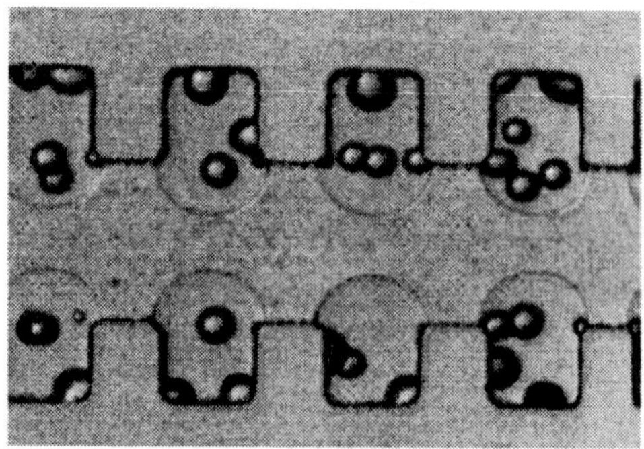

Figure 4 Condensation occurs in the air-filled cavities in a micro-channel

These droplets quickly grew to large drops that displaced air into the main section of the micro-fluidic channel. The cavities eventually filled with water, destroying the Cassie condition. We noticed that the drop growth rate increased with applied current, which agrees qualitatively with the trends we obtained on the droplet experiments on the super-hydrophobic surface. This process was highly repeatable. Moreover, the drops could be made to slowly shrink back and disappear when the current was turned off before the water filled in the cavities.

The phenomenon we observed in the micro-channel is analogy to the contact angle transition experienced for the water droplet on the cooled super-hydrophobic surface. This can be explained in terms of the mass transport. Because of the Joule heating effect, temperature of the buffer in the micro-channel is increased. Thus water in buffer evaporates into the air-filled cavities, and encounters the cool PDMS cavities walls which remain at a temperature close to room temperature. Thus, warm vapor evaporated from the liquid-vapor interface saturates the small cool cavities, and condensation quickly takes place. As long as the current is applied, the buffer is heated and the evaporating and condensing effect continue. The larger the applied current, the higher the temperature of the buffer, and the stronger the evaporating and condensing effects are. The result of this temperature gradient is a mass transfer from the liquid to the air-filled cavities.

Mass transfer in this manner can explain in instabilities in the Cassie condition often observed on super-hydrophobic surfaces. The temperature gradient between the liquid and the super-hydrophobic surface is responsible for accelerating the mass transfer. Once the condensation takes places, the transition from the Cassie state to the Wenzel state can be triggered.

4. CONCLUSIONS

The experimental results of a droplet placed on a controlled super-hydrophobic surface cooled underneath by a thermal electrical cooler demonstrate quick change in contact angles from the Cassie state composite contact to the Wenzel state wetting contact. The measured contact angles are consistent with the theoretical predictions of Cassie's and Wenzel's equations. The actual details of the transition are observed under a microscope through a specially designed one-dimensional micro-channel with concaved structures at the two sidewalls. The temperature-gradient-enhanced mass transfer results in a rapid condensation in the air-filled cavities, which is believed to be the possible mechanism to trig the instabilities of the energy states.

The phenomenon of mass transport into micro and nanocavities is important in understanding the nature of nano-structured super-hydrophobic surfaces. Cassie surfaces, which combine chemical and physical properties to produce the super-hydrophobic effect, are suspect to dynamic change based on thermal non-equilibrium states at the surface. Microfluidic designs intended to utilize super-hydrophobic or gas-filled walls will be affected by this phenomenon. As mass transfer occurs, trapped air bubbles will fill the main channel changing the flow dynamics, or even blocking flow completely. Furthermore, the microcavities will fill with a distilled solution that has different chemistry from the bulk solution, producing unplanned chemical conditions (e.g., concentration gradients) in the flow channel. Such changes to the local chemistry may be desirable—we recognize that control of this phenomenon may be put to good effect by creative design of microfluidic systems.

REFERENCES

1. T. Onda, S. Shibuichi, N. Satoh and K. Tsuji, Super-water-repellent fractal surfaces, *Langmuir* vol.12, pp.2125-2127, 1996.
2. C. Cottin-Bizonne, J. L. Barrat, L. Bocquet and E. Charlaix, Low-friction flows of liquid at nanopatterned interfaces, Nature Materials, vol.2, pp.237-240, 2003.
3. K. Watanabe, Y. Udagawa and H. Udagawa, Drag reduction of Newtonian fluid in a circular pipe with a highly water-repellent wall, *J. Fluid Mechanics*, vol.381, pp 225-238, 1999.
4. J. Bico, C. Marzolin and D. Quere, Pearl drops, *Europhys. Lett*, vol.52, pp.220-226, 1999.
5. N. A. Patankar, On the modeling of hydrophobic contact angles on rough surfaces, *Langmuir* vol.19, pp.1249-1253, 2003.

NANO2004-46084

ELECTROSTATIC EFFECT ON JUMP-TO-CONTACT AND ATOM TRANSFER BETWEEN AFM TIP AND SUBSTRATE

[1]M. C. Cheng, [1]J. L. Lai, [2]C. K. Sung and [3]H. M. Tai
[1]Graduate Students and [2]Professor
Department of Power Mechanical Engineering
National Tsing Hua University
Hsinchu, Taiwan 300, ROC
[2]cksung@pme.nthu.edu.tw
[3]Associate Researcher
Center for Measurement Standards, ITRI

ABSTRACTX

Dynamics of atoms around the interface of tip and substrate has close relationship to adhesive forces. This paper investigates the interactive behaviors by utilizing molecular dynamic (MD) simulation with the consideration of external applied electrostatic (ES) field between the atomic force microscopy (AFM) tip and substrate, which are both coated with gold. Morse potential was employed herein for modeling the energy between the atoms on the tip and substrate. The simulation model is composed of the gold-coated parts of the AFM tip and substrate, which are connected to the rest parts of the AFM tip and substrate, respectively, by the use of boundary conditions. The gold atoms were arranged in order according to face centric cubic (FCC) structure. The tip with the pyramidal shape was formed from 664 gold atoms while the substrate with cuboid shape was constructed from 2400 gold atoms. Simulation results show that both jump-to-contact interactions and the atom transfer increase as a result of enhancing the bias voltage between the tip and the substrate. The results agree well with experimental observation and also have the same trend as the experiment.

1. NTRODUCTION

Atom transfer is one of the techniques for the fabrication of nanostructures. Since the electrostatic field employed in the nano-fabrication can affect the interactive forces between the tool and workpiece, application of ES field is a possible method for controlling the manufacturing process precisely. This paper integrates an AFM tip as the mechanical tool and a substrate as workpiece with the external applied ES field as a model for exploring the ES effect on the jump-to-contact and atom transfer phenomena. In this article, some boundary conditions and algorithms are included for describing the ES interaction between AFM tip and substrate.

From the physical prospect, van der Waals force, ES force, and force caused by liquid bridge can be used to determine the force interactions in the AFM system. To investigate the interactive behavior between the tip and the substrate, the three-dimensional (3D) Molecular Dynamics (MD) simulation has been proved to be an effective tool. Landman et al. [8,9] first studied the nanomechanics and dynamics between the tip and substrate by MD simulation. Their studies successfully showed the dynamics of atoms around the interface of tip and a substrate. Fang et al. [10] demonstrated a scribing process on

the sample by MD simulation and the results fit in with their experiment. Miesbauer et al. [11] simulated the adhesion, jump-to-contact, and indentation between the NaCl crystals. Komanduri et al. [12] simulated a nanoscratching process by MD and they exhibited many quantitative data of the atomic scale friction. The main objective of this study is to utilize MD simulation to investigate the interactive behaviors between the tip and the substrate with the consideration of external applied electrostatic forces

In this simulation, the Morse potential [2] is used to study the force interaction of the AFM system when the tip touches and leaves the substrate. Finally, the simulation results are compared with the AFM experimental results.

2. SIMULATION MODEL

The MD simulation model is a tip-substrate configuration [3, 4]. As shown in Fig. 1, this model contains a part of the AFM tip, which is coated with gold and connected to the main part of the tip by the use of boundary conditions, and substrate. The gold atoms are arranged in order according to face centric cubic (FCC) structure. The tip with the pyramidal shape is formed from 664 gold atoms and the substrate with the cuboid shape from 2400 gold atoms. The lengths in X, Y, and Z coordinates of the substrate are 10, 10 and 6 times lattice constant, respectively. The distance between the bottom of tip and top surface of the substrate is 3.5 times lattice constant.

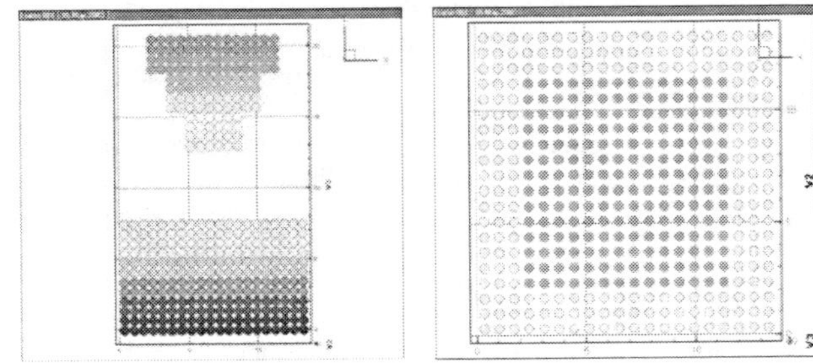

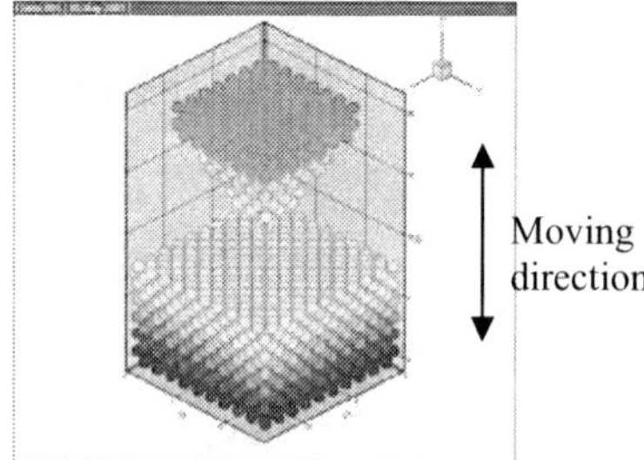

Figure 1 The simulation model

The motions of free atoms are governed by the potential function and the kinematic equation. The empirical potential function used for the simulation is a pairwise sum of the Morse potential:

$$u\left(r_{ij}\right) = D\left\{ e^{-2\alpha\left(r_{ij}-r_0\right)} - 2e^{-\alpha\left(r_{ij}-r_0\right)} \right\} \qquad (1)$$

where $u(r_{ij})$ is a pair potential energy function, the cohesive energy D, the equilibrium distance r_0, and the elastic parameter α. Their related parameters are shown in Table 1 [13]. The force on atom i resulting from interaction of all the other atoms can be derived from above potential function Eq.(1) such that:

$$F\left(r_{ij}\right) = -\frac{du\left(r_{ij}\right)}{dr} = 2\alpha D\left[e^{-2\alpha\left(r_{ij}-r_0\right)} - e^{-\alpha\left(r_{ij}-r_0\right)} \right] \qquad (2)$$

where $F(r_{ij})$ is the reactive force on atom i.

Table 1. Parameters in the Morse potential for gold [13]

Parameter	Value
D	0.56 (eV)
α	1.637 (Å^{-1})
r_0	2.922 (Å)

The Morse potential was utilized here due to its computational inexpensive, simplicity and accuracy and it has been used previously in several studies [10, 12, 13].

In the simulation, the X and the Y directions of the substrate are set as the periodic boundary condition, and the top layer of the tip and the bottom layer of the substrate are constrained to be the fixed boundary condition, they are assigned to control the motion of the tip, maintaining the position of the substrate. From physical point of view, the atoms in the fixed layers should be absolute zero in temperature, therefore, there should exist some atoms to dissipate energy and control state of the system, these atoms are called thermostatic atoms and the layers between fixed and periodic boundary condition are set to be thermal control boundary condition. The others are set to be free boundary condition. The initial position of tip is arranged in pyramidal shape and the substrate is in cuboid shape. Initial velocities are assigned from the Maxwell distribution, and the magnitudes are

scaled to the temperature of the system we desired, the method of controlling the total kinetic energy according to scaling technique [6]:

$$V_i^{new} = V_i \cdot \sqrt{\frac{E_{kd}}{E_{ka}}} \quad , \quad E_{kd} = \frac{3}{2}NkT_d \qquad (3)$$

In Eq. 3, T_d is the desirable temperature; E_{kd} is the kinetic energy; k is the Boltzmann's constant, and N is the number of atoms in thermal layers.

Without loss of generality, some assumptions are made for facilitating the simulation. Due to gold's excellent conductivities, the charges are assumed to be well-distributed on the surfaces of the tip and the substrate. The tip of the model is assumed to be symmetric in the X and Y directions, because of its relatively small size of the tip. Only the electrostatic force interaction in the Z direction is considered and the periodic boundary effect in the X and Y directions for the sample are neglected. The charge distribution is assumed to be similar to the parallel capacitor, which means that the charges of the tip atoms and the sample atoms are all the same. The ionization and the electromagnetic effect in the charge boundary condition are also neglected. In the simulation, we set the voltage V as the control parameter, and the charges of the atoms are calculated from the following equation.

$$V = \frac{Q_{Tip}Q_{Substrate}}{4\pi\varepsilon_0 Z_{Tip-Substrate}} \quad , \quad Q_{Tip} = Q_{Substrate} \qquad (4)$$

$$q_{i,Tip} = -\frac{\sqrt{4\pi\varepsilon_0 Z_{Tip-Substrate}}V}{N_{Tip}} \qquad (5)$$

$$q_{i,Substrate} = \frac{\sqrt{4\pi\varepsilon_0 Z_{Tip-Substrate}}V}{N_{Substrate}} \qquad (6)$$

where N_{Tip} is the number of the charged atoms in the tip model and it is similar to $N_{Substrate}$. The electrostatic force between atoms is

$$F_{ij} = \frac{q_i q_j}{4\pi\varepsilon_0 r_{ij}^2} \qquad (7)$$

There is no clear and definite surface in the nano-scale size because the atomic nucleus is surrounded by the electric cloud and the distance between atoms changes when atoms approach to each other. For solving this problem, the displacement between the tip and sample surface is assumed as the relative motion of the fixed boundaries defined for the tip-sample model. In the MD simulation, all the motion of the atoms should follow the Newton's second law of motion. However, the initial displacements and velocities are determined independently and the fifth Gear's predictor-corrector algorithm [6] is employed for time integration of motion. In addition, Verlet neighbor lists scheme [6] is utilized for improving computation efficiency.

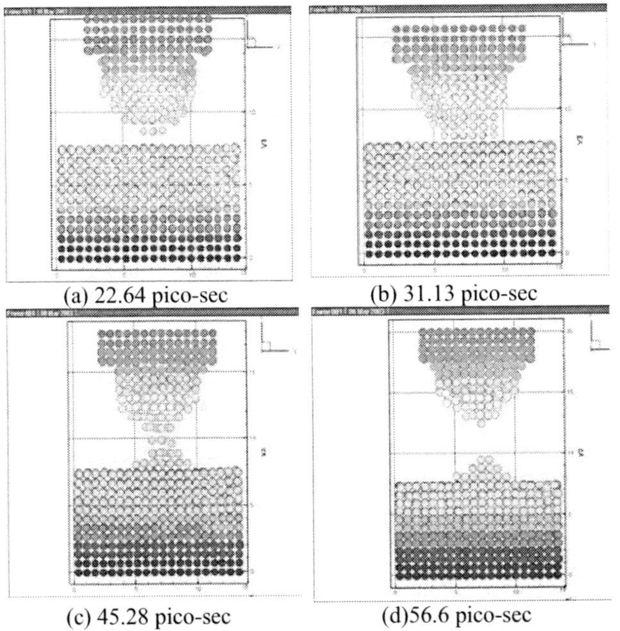

(a) 22.64 pico-sec   (b) 31.13 pico-sec

(c) 45.28 pico-sec   (d)56.6 pico-sec

Figure 2 The tip motion in the simulation where the voltage between the tip and the substrate is 0V

3. Results and Discussion

In this simulation, the motion of fixed layer of the tip is controlled by lowering it down to touch the substrate surface (time step: 1-9500) and stopping for a while (time step: 9500-11500) to dissipate the fluctuations caused by the mechanical contact, then lifting up to leave the substrate (time step: 11500-22000), as shown in Fig. 3. The displacement of the tip is about 2.08nm and the time step for the simulation is selected to be 2.83 fsec. The voltage is considered as the control parameter to observe the interactive force and the atom transfer between tip and substrate. Herein, the tip is set to be the negative polarity. Fig. 4 and 5 are force curves that respectively illustrate phenomena such as jump-to-contact, necking and nanostructure formation under different bias voltage, 0 and 8V.

3-1 Jump-to-contact

Due to the interactions between molecules, the molecules on the bottom layers of tip and that on the top layers of substrate tend to attract each other when the distance between them is getting shorter. This behavior is generally referred to as the jump-to-contact phenomenon at atomistic level by Landman et al. [8, 9] in the model of Ni and Au atoms. In the simulation, for gold atoms, this phenomenon is more apparently. In addition, the bias voltage is applied to the tip and the substrate, so, from Figure 5, one can observe the time step when jump-to-contact occurs, which is earlier than Figure 4 and the force is larger, too.

3-2 Necking

In the tip ascending process of the simulation, owing to the adhesive force between the tip and the substrate, the necking is formed. In addition, when the tip lifts up to leave the substrate, the thickness changes; if we stop the process, we can infer to as a method to create "nanowire."

3-3 Nanostructure formation

When the tip continues to ascend, the neck becomes thinner and thinner, then, it breaks into two parts. The one that stays on the substrate is nanostructure; sometimes we call it island structure. The size of the island (defined by the number of atoms stayed on the substrate) will be different, if we apply different bias voltages, as shown in Figure 6.

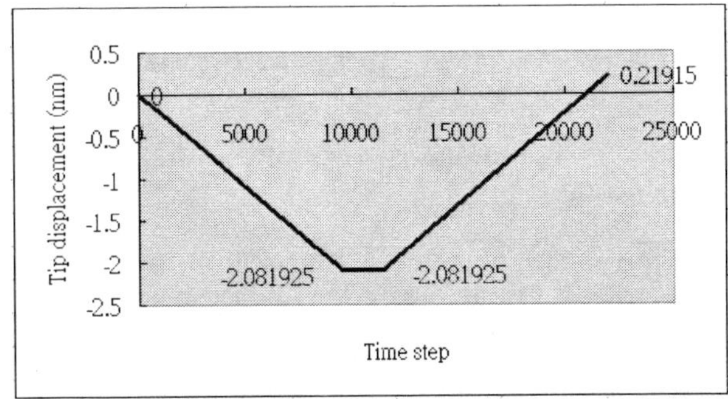

Figure 3 The displacement of the tip in the MD simulation.

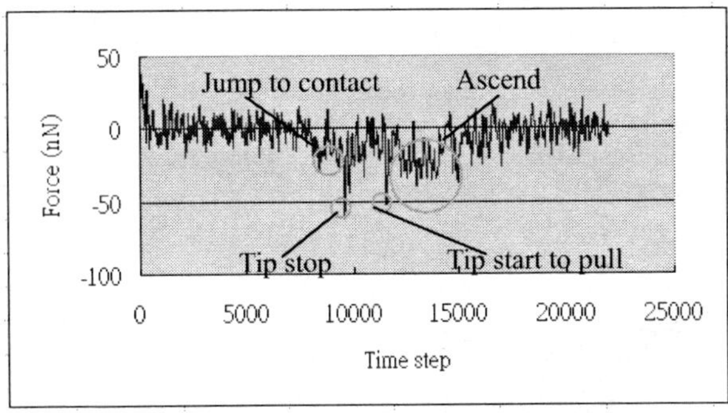

Figure 4 The tip force curve when the voltage between the tip and the substrate is 0V

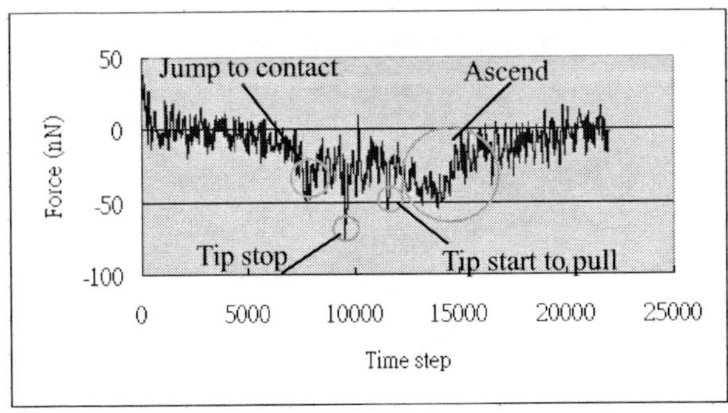

Figure 5 The tip force curve when the voltage between the tip and the substrate is 8V

81

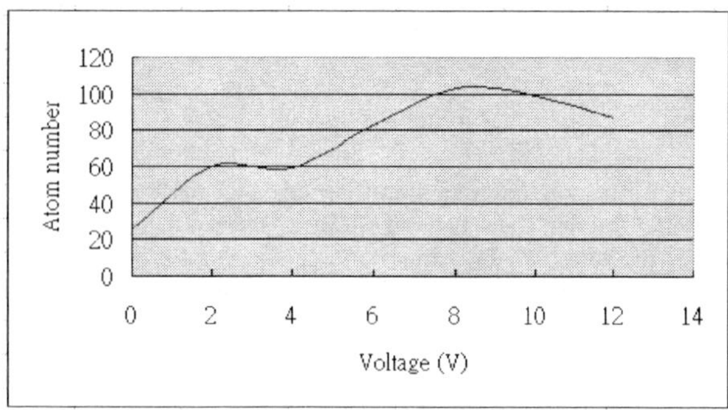

Figure 6 The relation curve between the number of the atoms which leave the tip and stick on the substrate after the contact and the voltage.

4. Conclusions

An approach by using the three dimensional molecular dynamic simulation is presented to evaluate the effect of bias voltage on the interatomic forces between the tip and substrate at atomic scale. The simulation results indicate that the jump-to-contact phenomenon occurs as the tip approaches the substrate. The necking phenomenon appears as the tip lifts up after the tip contact the substrate owing to the adhesion between the tip and the substrate, then the gold atoms transfers from the tip to the substrate to form the nanostructure. When we applied the bias voltage on the tip and the substrate, we can observe that the larger the voltage between the tip and the substrate is, the greater the jump-to-contact force and the pull-off force are, besides, the more the gold atoms are transferred from the tip to the substrate.

REFERENCES

[1] M. Sitti, "Survey of Nanomanipulation Systems," *IEEE-NANO 2001*, S2.2 Nanomanipulation I (Special session).

[2] L. A. Girifalco and V. G. Weizer, "Application of the Morse Potential Function to Cubic Metals," Phys. Rev., Vol. 114, No. 3, 1959.

[3] M. R. Sorensen, K. W. Jacobsen, and P. Stoltze, "Simulations of atomic-scale sliding friction", Physical Review B, Vol. 53, No. 4, pp. 2101-2113, 15 January 1996.

[4] H. Rafii-Tabar, "Modelling the nano-scale phenomena in condensed matter physics via computer-based numerical simulations", Physics Reports, 325, pp.239-310, 2000.

[5] W. Eckstein, "Computer simulation of ion-solid interaction," Spring-Verlag, 1991.

[6] J. M. Haile, "Molecular Dynamics Simulation-Elementary Methods," Clemson University, Clemson, South Carolina.

[7] T. T. Tsong, "Effects of electric field in atom manipulations," Physical Review B 44, 13703, 1991.

[8] U. Landman, W. D. Luedtke, N. A. Burnham, and R. J. Colton, "Atomistic mechanisms and dynamics of adhesion, nanoindentation, and fracture," Science, Vol.248, pp. 454-461, 1990.

[9] U. Landman and W. D. Luedtke, "Nanomechanics and dynamics of tip-sample interaction," J. Vac. Sci. Technol. B, Vol. 9(2), pp. 414-422, 1991

[10] T. H. Fang, C. I. Weng, and J. G. Chang, "Molecular dynamics simulation of nano-lithography process using atomic force microscopy," *Surf. Sci.*, Vol. 501, pp. 138-147, 2002.

[11] O. Miesbauer, M. Götzinger, and W. Peukert, "Molucular dynamics simulations of the contact between two NaCl nano-crystals: adhesion, jump-to-contact and indentation," *Nanotech.*, Vol.14, pp. 371-376, 2003

[12] R. Komanduri, N. Chandrasekaran, and L. M. Raff, "Molecular dynamics simulation of atomic-scale friction," *Phys. Rev. B,* Vol. 61, pp.14007-14019, 2000

[13] K. Maekawa and A. Itoh, "Friction and tool wear in nano-scale maching – a molecular dynamics approach," Wear, Vol. 188, pp. 115-122, 19

NANO2004-46085

Impact of Electron-Phonon Non-equilibrium on Heat Dissipation of Metallic Nano-Particles Suspended in Dielectric Media

Y. Sungtaek Ju
Department of Mechanical and Aerospace Engineering
UCLA
Los Angeles, CA 90095-1597
e-mail) just@seas.ucla.edu
tel) (310) 825-0985

Controlled heating of nano-particles is a key enabling technology for various nano-manufacturing and biomedical applications. They include controlled fabrication of nano-particles,[1] and targeted heating of biological molecules and cells for research as well as therapeutic purposes.[2] Recent studies also demonstrated improved heat transfer properties of colloids of nano-particles.[3]

A recent experimental study[4] measured cooling time constants of gold nano-particles suspended in water and subjected to sub-picosecond pulsed laser heating. While the continuum heat diffusion model correctly captured the particle size dependence, the absolute values of the time constant were significantly underpredicted. The discrepancy can be partly explained by finite thermal resistance at the interface between nano-particles and media, which impedes heat conduction cooling. Including the thermal interface resistance in the continuum model does result in increased time constant but the agreement with the data is still poor. The dashed and dotted lines in Figure 1 represent the time constants predicted by assuming two different values of the thermal interface resistance, 0.5 and 1 x10^{-8} m^2 K/W. The thermal interface resistance apparently increases with increasing particle size.

The above prediction implicitly assumed that electrons and phonons within Au nano-particles are in equilibrium with each other. This is normally justifiable because electron – phonon coupling time (~ ps) is a lot shorter than the measured cooling time constants. If heat transfer to the surrounding media is significant, however, electrons and phonons may exhibit *spatial* as opposed to temporal non-equilibrium. This is because electrons interact indirectly with atomic vibrations in the adjacent dielectric medium. Energy transfer across the interface is mediated by phonons, whose contribution to heat conduction is often ignored in pure bulk metals.

We use the two-temperature heat diffusion model[5] to study the possible impact of phonon-electron non-equilibrium on cooling dynamics of nano-particles:

$$C_e \frac{\partial T_e}{\partial t} = \nabla \bullet \left(k_e \nabla T_e \right) - G(T_e - T_{ph}) + q''' \tag{1}$$

$$C_{ph} \frac{\partial T_{ph}}{\partial t} = \nabla \bullet \left(k_{ph} \nabla T_{ph} \right) + G(T_e - T_{ph}) \tag{2}$$

We solve Equations (1-2) numerically to compute temporal variations in electron temperature within nano-particles. The predicted time constants are shown in Fig. 1 and agree well with the data for all particle sizes considered. Based on solutions to a related problem in the planar 1-D geometry, we define the electron-phonon coupling length $\delta = [\, k_e\, k_{ph} / (k_e + k_{ph})\, G\,]^{1/2}$ as the characteristic interaction distance required for electrons and phonons to reach equilibrium. For particles whose diameter is much larger than this length, electron-phonon non-equilibrium contributes extra resistance of the order of δ/k_{ph}.

The present analysis suggests that non-equilibrium between electrons and phonons in nano-particles has significant impact on the relaxation of electron temperature even on timescales greater than the electron-phonon coupling time. The prediction of the continuum model agrees well with the experimental data when both the non-equilibrium effect and the finite thermal interface resistance are taken into account.

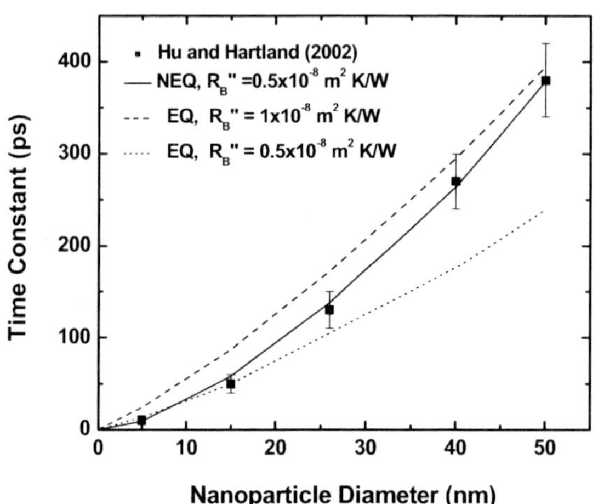

Figure 1: The characteristic time constants of electron cooling in Au nano-particles as a function of particle size.

References

[1] .H. Kurita, A. Takami, and S. Koda, Appl. Phys. Lett. 72, 789 (1998).

[2] K. Hamad-Schifferli, J. J. Schwartz, A. T. Santos, S. Zhang, and J. M. Jacobson, Nature 415, 152 (2002).

[3] J. A. Eastman, S. U. S. Choi, S. Li, W. Yu, and L. J. Thompson, Appl. Phys. Lett. 78, 718 (2000).

[4] M. Hu and G. V. Hartland, J. Phys. Chem. B 106, 7029 (2002)

[5] S. I. Anisimov, B. L. Kapeliovich, and P. L. Perelman, Sov. Phys. JETP 39, 375 (1974).

NANO2004-46090

SURFACE TENSION PREDICTION FROM DENSITY PROFILE INFORMATION BY POLYATOMIC MOLECULAR DYNAMICS SIMULATIONS

Aaron P. Wemhoff and Van P. Carey
Department of Mechanical Engineering
University of California
Berkeley, CA 94720-1174, USA
wemhoff@me.berkeley.edu
vcarey@me.berkeley.edu

Surface tension prediction of liquid-vapor interfaces of polyatomic fluids using traditional methods in molecular dynamics simulations has shown to be difficult due to the requirement of evaluating complex intermolecular potentials even though these methods provide accurate predictions. In addition, the traditional methods may only be performed during a simulation run. However, analytical techniques have recently been developed that determine surface tension by using the characteristics of the density profile of the interfacial region between the bulk liquid and vapor regions. Since these characteristics are a standard result of many liquid-vapor interfacial region simulations, these data may be used in a post-simulation analysis. One such method, excess free density integration (EFEDI), provides results from the post-simulation analysis, but the expansion from monatomic to polyatomic fluids is not straightforward [1]. A more general and powerful approach to surface tension involves the application of a Redlich-Kwong-based mean-field theory [2], which has resulted in a single equation linking the surface tension of a fluid, σ_{lv}, with the density gradient at the center of the interfacial region,

$$\sigma_{lv} = 0.1065(1 - T/T_c)^{-0.34} L_i^2 \left(\frac{d\hat{\rho}}{dz}\right)_{z=0} \frac{a_{R0}N_A^2}{b_R N_A T^{1/2}} \ln\left(\frac{1 + \hat{\rho}_l b_R N_A}{1 + \hat{\rho}_v b_R N_A}\right) \quad (1)$$

where z is the position normal to the interfacial region and is zero at its center, $\hat{\rho}_l$ and $\hat{\rho}_v$ are the liquid and vapor molar densities, respectively, T_C is the critical temperature, N_A is Avogadro's number, L_i is a characteristic length given by

$$L_i = \left(\frac{k_B T_C}{P_C}\right)^{1/3} \quad (2)$$

and a_{R0} and b_R are the coefficients in the Redlich-Kwong equation of state,

$$P = \frac{N k_B T}{V - b_R N} - \frac{a_{R0} N^2}{T^{1/2} V(V + b_R N)} \quad (3)$$

Furthermore, P_C is the critical pressure for the fluid. Reference [2] shows that the relation provided by Equation 1 provides a approximate prediction of surface tension for argon fluid using data from molecular dynamics simulations. The derivation of Equation 1 is based on the assumption that the density profile in the interfacial region follows

$$\frac{\hat{\rho} - \hat{\rho}_v}{\hat{\rho}_l - \hat{\rho}_v} = \frac{1}{e^{4z/\delta z_i} + 1} \quad (4)$$

where δz_i is the interfacial region thickness,. Note that Equation 4 is more commonly expressed in the equivalent form

$$\hat{\rho}(z) = \frac{1}{2}(\hat{\rho}_l + \hat{\rho}_v) - \frac{1}{2}(\hat{\rho}_l - \hat{\rho}_v)\tanh\left(\frac{2z}{\delta z_i}\right) \quad (5)$$

Wemhoff and Carey [1] have recommended the use the fit curve relation given by Equation 5 for the liquid-vapor interfacial region of a diatomic nitrogen system. Therefore, Equation 1 may be used to predict the surface tension for diatomic nitrogen at various temperatures.

The diatomic nitrogen system simulated in this study initialized 864 nitrogen molecules in a face-centered-cubic lattice between two sparse vapor regions of less than 20 molecules. All molecules were initialized with translational and rotational velocities corresponding to the Boltzmann translational and rotational distributions for a low initial reduced temperature ($T_r = T/T_C = 0.1$). The system was allowed to equilibrate to the desired equilibrium temperature via velocity rescaling for 50,000 time steps of 1.0 fs, and density profile information was collected in 100 bins along the z-axis for an additional 30,000 to 50,000 steps as an NVE-type system. The molecules interacted via the two-center Lennard-Jones potential, where each atom in the nitrogen molecule interacted with all atoms in other nitrogen molecules via the simple Lennard-Jones 6-12 potential,

$$\phi_{ij}(r) = 4\varepsilon\left[\left(\frac{\sigma}{r_{ij}}\right)^{12} - \left(\frac{\sigma}{r_{ij}}\right)^{6}\right] \tag{6}$$

where r_{ij} is the distance between atom i and atom j, and ε and σ are given as $36.673k_B$ K and 0.33078 nm, respectively. Interactions were truncated at 3.0σ, periodic boundary conditions on all sides of the domain were imposed, and the rigid bond length between atoms in the nitrogen molecule was 0.3292σ and maintained by the RATTLE algorithm [3]. Figure 1 shows an equilibrium molecular configuration for a diatomic nitrogen liquid-vapor system.

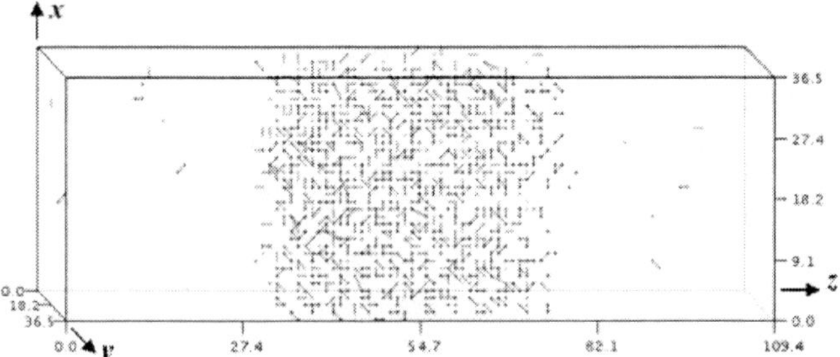

Figure 1. Equilibrium snapshot of diatomic nitrogen liquid-vapor system, $T_r = 0.60$.

The gradient used in Equation 1 was taken as the average of the values found from both sides of the film, and the gradient value for a given interfacial region was approximated by using the difference of density values of the bins closest to the center of the interfacial region as shown in Figure 2. Figure 3 shows the dimensionless surface tension predicted by Equation 1 for a variety of reduced temperatures, where the dimensionless surface tension is given as $\gamma/(P_C L_i)$. Also shown in the figure are values predicted by the EFEDI analysis [1] and recommended values by ASHRAE [4]. Figure 3 shows that the predictions made by Equation 1 overestimate the surface tension of the fluid by approximately 40%. This fact is not surprising given that approximately the same degree of overestimation is given in their prediction of surface tension for argon [2]. Therefore, the density gradient relation appears to be as effective for a diatomic nonpolar fluid as a monatomic fluid. The method presented here may also extended to analysis of other fluids.

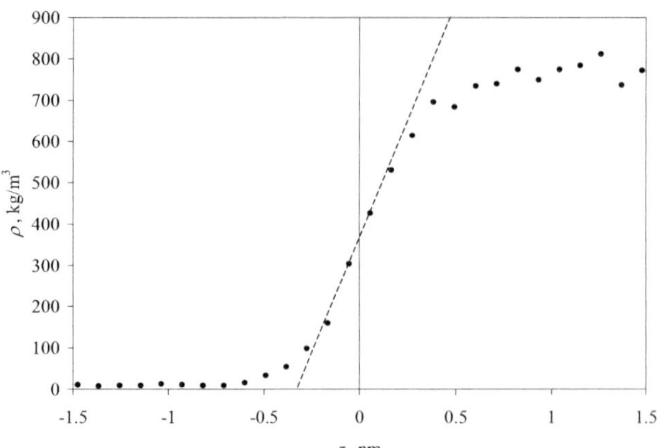

Figure 2. Approximation of the density gradient from collected bin density information. Results shown for $T_r = 0.60$.

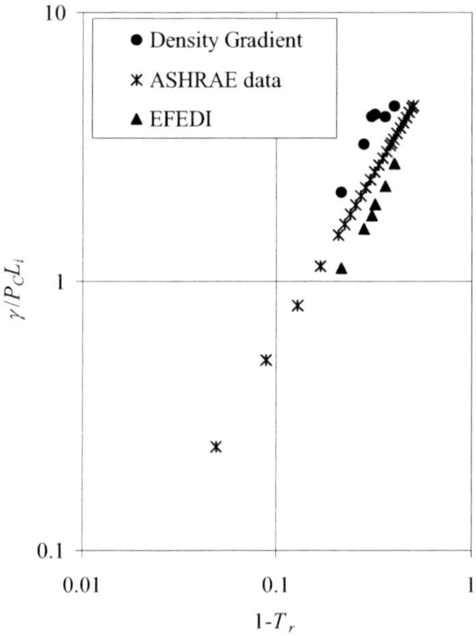

Figure 3. Predicted surface tension values using the density gradient relation of Equation 1 compared to values by Wemhoff and Carey [1] and recommended values by ASHRAE [4]

REFERENCES

[1] Wemhoff, A.P., and Carey, V.P., 2004, Surface Tension Evaluation via Thermodynamic Analysis of Statistical Data from Molecular Dynamics Simulations, *Proc. 2004 Heat Trans./Fluids Engineering Summer Conf.*, Charlotte, NC.

[2] Carey, V.P., and Wemhoff, A.P., 2004, Relationships Among Liquid-Vapor Interfacial Region Properties – Predictions of a Thermodynamic Model, *Int. J. Thermophysics*, in press.

[3] Andersen, H. C., 1983, Rattle: a 'Velocity' Version of the Shake Algorithm for Molecular Dynamics Calculations, *J. Comput. Phys.*, Vol. 52, pp. 24-34.

[4] American Society of Heating, Refrigerating, and Air-conditioning Engineers, 2001, *ASHRAE Fundamentals Handbook*, ASHRAE, Atlanta, GA.

NANO2004-46091

MAGNETICALLY ASSEMBLED 3-D MESOSCOPIC PATTERNS USING SUSPENSION OF SUPERPARAMAGNETIC NANOPARTICLES

Ashok Sinha
Department of Engineering Science & Mechanics
Virginia Tech, Blacksburg, VA

Ranjan Ganguly[1]
Mechanical and Industrial Engineering Department
University of Illinois,Chicago, IL

Ishwar K. Puri
Department of Engineering Science & Mechanics
Virginia Tech, Blacksburg, VA

Keywords: Ferrofluid; Self assembly;

ABSTRACT

A self assembly process is characterized by the spontaneous and ordered aggregation of similar components. At the nanoscale, these constituents can be colloids, or other nanoparticles that combine to form structures of meso- and macroscopic dimensions. Self assembly is most evident during the growth of biological structures. Due to its natural elegance, there is considerable interest in investigating similar methods in other fields of sciences. In principle, the process of dynamic self assembly is characterized by a competition between at least two forces – one attractive and the other repulsive – in thermodynimacally nonequilibrium systems. Several researchers have described self assembly in configurations where biological [1] or chemical [2] mediation, electrostatic [3], capillary [4] or fluid dynamic [5] forces have provided a motive potential.

The self assembly of superparamagnetic nanoparticles is a promising technique for producing high density magnetic data storage devices and three dimensional memory chips. Magnetically assembled mesoscopic structures will have promising application in surface roughening (for heat and mass transfer augmentation) or even producing microelectronic devices. The same strategy can be extended to produce microscale structures, too.

Ferrofluids are colloidal suspensions of single domain (i.e. superparamagnetic) nanoparticles that are typically of the order of 10 nm in diameter. The nanoparticles are coated with adsorbed surfactant layers to prevent particle agglomeration due to the van der Waals forces and dipole-dipole interactions among them. Consequently, a ferrofluid can often be considered as a single homogeneous liquid [6]. The nanoparticles in the ferrofluid can be collected into aggregates by applying external magnetic fields [6,7] or they can form patterns on flat substrates [8, 9]. For instance, a thin ferromagnetic film can be patterned into isolated islands and used to direct the assembly of superparamagnetic colloidal particles into two-dimensional arrays [10]. However, previous research has focused on the self assembly of two-dimensional patterns rather than of complex three-dimensional architectures.

Our study focuses on forming permanent 3-D mesoscopic structures. When a sessile drop of ferrofluid placed on a flat substrate is subjected to a spatially varying magnetic field, the interactions between gravity, surface tension and the magnetic force create different shapes [11]. Figure 1(a) presents an image of a mesoscale 3-D structure produced by a hydrocarbon oil-based EFH 1 ferrofluid (FerroTec USA, NH) that was placed on a placed on a wetting glass surface under the influence of the nonuniform magnetic field. The height, shape and orientation of such microliter droplets can be adjusted by altering the nature of the magnetic field, as shown in Fig. 1(b) for droplets on a wetting glass

---

[1] Corresponding author: rgangu2@uic.edu; Also: Power Engineering Department, Jadavpur University, Kolkata, India

surface under the influence of the nonuniform magnetic field. The height, shape and orientation of such microliter droplets can be adjusted by altering the nature of the magnetic field, as shown in Fig. 1(b). Although the 3-D structures thus formed are not permanent (i.e. they disappear when the magnetic field is withdrawn), the ferrofluid can be premixed with suitable curing agents. Once the mixture is cured in the magnetically assembled position, the structures are permanent.

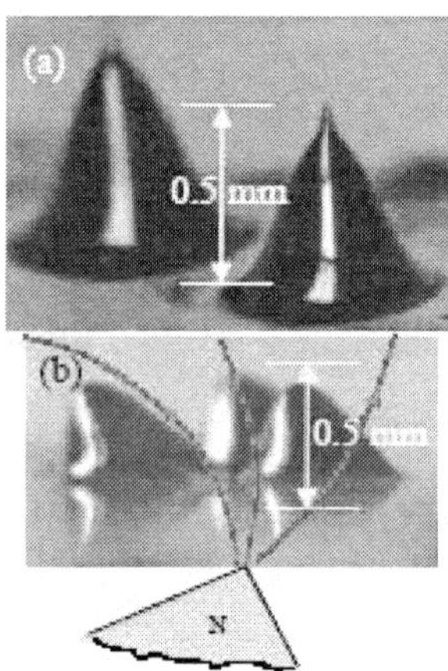

Fig. 1 (a) Three-dimensional ferrofluid structures produced on a flat substrate using a bar magnet; and (b) varying orientations of these structures with respect to the magnetic lines of force.

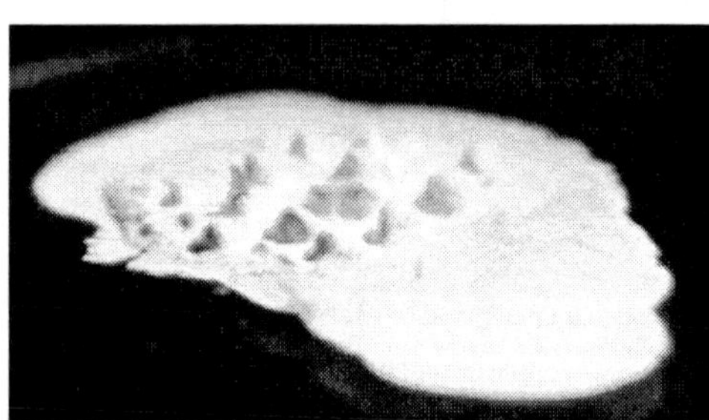

Fig. 2: Permanent structures formed by premixing a cementing agent with ferrofluid and curing under the magnetic influence

Figure 2 shows a permanent structure formed when a cementing agent is premixed with the ferrofluid. The dimensions of the 3-D conical structures are of the same order of the ones shown in Fig. 1. However, the choice of the cementing agent and its percentage composition in the mixture is important. Since the curing agent is nonmagnetic, the magnetic force in the mixture liquid is lower than the same on a pure ferrofluid. This is apparent from the relatively blunt peaks of the structure.

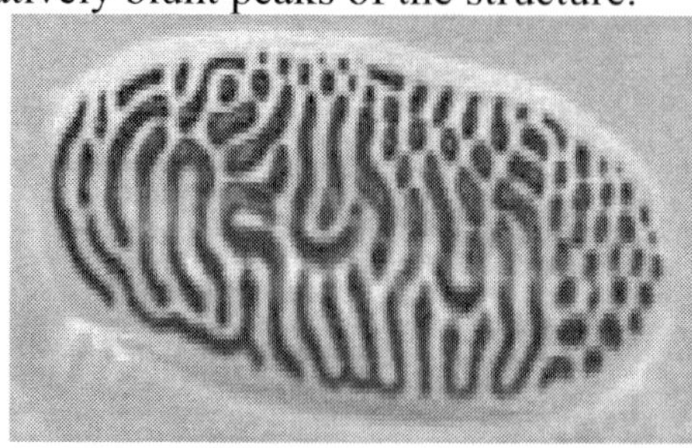

Fig. 3: Labyrinthine instability pattern shown by ferrofluid

When a thin film of ferrofluid shares an interface with another immiscible liquid and a strong uniform magnetic field is applied, labyrinthine patterns are formed as shown in Fig. 3. Although, this pattern is observed in thin films only [6], it can have wide application. If permanent and well-controlled labyrinthine structures can be generated using the ferrofluid- curing agent mixture (as discussed before) this can potentially revolutionize the microfabrication processes for microchannels, miniaturized catalytic converter beds, and spiral combustors.

REFERENCES

[1] P. K. Maini, K.J. Painter, and H.N.P. Chau, J. Chem. Soc., Faraday Transaction, **93**, 3601 (1997).
[2] R. Kapral, Physica D **86**, 149 (1995).
[3] B.A. Grzybowsky, A. Winkleman, J. A Wiles, Y. Brumer, and G. M. Whitesides, Nat. Mat. **2**, 241 (2003).
[4] N. Bowden, A. Terfort, J. Carbeck, and G. M. Whitesides, Science **276**, 233 (1997).
[5] H-J. J. Yeh and J. S. Smith, IEEE PhotonicsTech. Lett. **6**, 706 (1994).
[6] R.E. Rosensweig, *Ferrohydrodynamics* (Dover Publications Inc., 1997).
[7] S. Odenbach, Colloids Surfaces A **217**, 171 (2003).
[8] R. E. Goldstein, D. P. Jackson, and S. A. Langer, J. Magn. Magn. Mater. **122**, 267 (1993).
[9] I. J. Jang, H. E. Horng, Y. C. Chiou, C-Y. Hong, J. M. Yu, and H. C. Yang, J. Magn. Magn. Mater. **201**, 317 (1999).
[10] B. Yellen, G. Friedman, and A. Feinerman, J. Appl. Phys. **93**, 7331 (2003).
[11] A.G. Boudouvis and L. E. Scriven, J. Magn. Magn. Mater. **122**, 254 (1993).

NANO2004-46092

Synthesis of Polycrystalline Zeolite Films and Thermal Conductivity Measurements by a 3-Omega Method

Yeny Hudiono
School of Chemical & Biomolecular Engineering
Georgia Institute of Technology, Atlanta, GA

Adam Christensen
School of Mechanical Engineering
Georgia Institute of Technology, Atlanta, GA

Weontae Oh
School of Chemical & Biomolecular Engineering
Georgia Institute of Technology, Atlanta, GA

Samuel Graham
School of Mechanical Engineering
Georgia Institute of Technology, Atlanta, GA

Sankar Nair
School of Chemical & Biomolecular Engineering
Georgia Institute of Technology, Atlanta, GA

ABSTRACT

Zeolites (nanoporous crystalline aluminosilicates) have important applications in catalysis and separations [1,2], and are also being considered for adsorption-based heating and cooling systems [3]. We are investigating the use of zeolite films in the fabrication of more efficient adsorption heat pumping and refrigeration systems that use water vapor as the working fluid. The thermal conductivity of the adsorbent is an important property affecting heat transfer in an adsorption heat pumping system. There are few reports (e.g. [4]) of the thermal conductivity of zeolites, which is measured by compacting the zeolite powder into a disk sample and using a model to extract the 'intrinsic' thermal properties. Another approach [5] relies on molecular dynamics simulation using parameterized force fields to predict the intrinsic thermal conductivity.

In this study, polycrystalline zeolite thin films [6] are synthesized and used to measure the thermal conductivity of zeolites by a 3-omega method reported recently [7]. By using a zeolite film with intergrown crystals rather than a compacted powder sample, thermal conductivities close to that of a single crystal can be obtained. The nanoporous zeolite MFI with pore size of approximately 0.6 nm was grown hydrothermally into oriented films (*ca.* 25 microns thin) by the secondary (seeded) growth method. The secondary growth consists of two steps. Firstly, zeolite nanocrystals (~ 100 nm in size) are deposited onto a glass substrate by dip coating from a zeolite suspension, at the rate of 2 cm/hour in order to obtain a uniform seed layer. The coated substrate is air dried for approximately 24 hours, and then is placed on the bottom of a Teflon autoclave, leaning on the wall of the autoclave. The membrane growth solution

containing silica and alumina suspension is mixed at room temperature and poured into the autoclave. The autoclave is

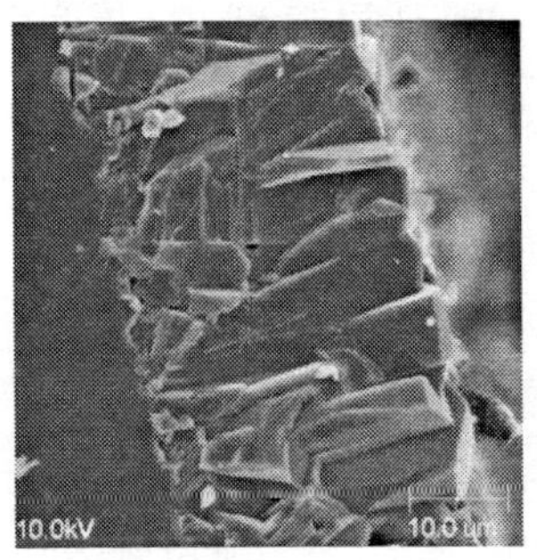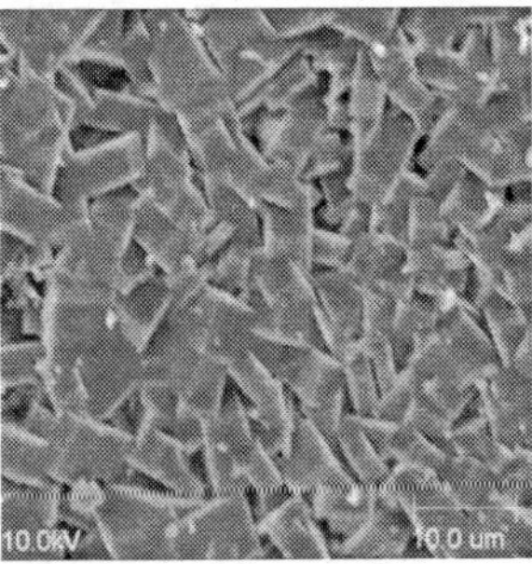
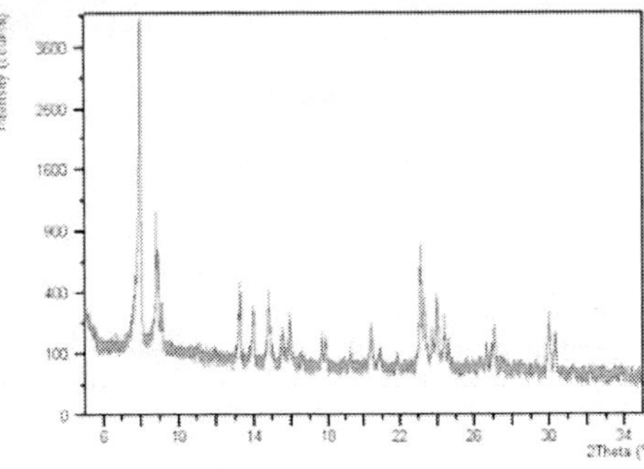

Fig. 1. SEM images of (top left) cross-section, (top right) surface of calcined MFI membrane after 24 hr of hydrothermal treatment, and (bottom) X-ray diffraction pattern of MFI membrane.

sealed and heated at 453 K for 1-3 days. The membranes are rinsed with hot water and then dried. Some of the membranes are calcined in order to remove the organic template. SEM images of calcined MFI membrane and the X-Ray diffraction show an oriented columnar structure, as shown in Figure 1.

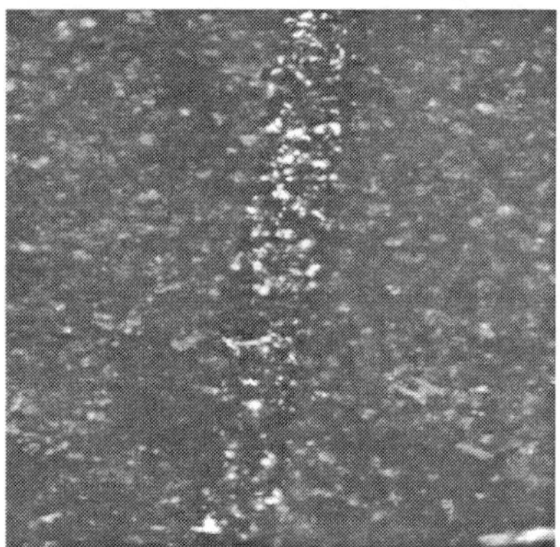

Fig. 2: Optical microscope image of copper line heater on the zeolite fillm.

The thermal conductivities of these thin films have been measured as a function of temperature using a 3ω method [7]. A thin copper strip deposited on the zeolite membrane (Figure 2), is used to flow an electrical current which is used to heat the zeolite film. The width and the length of the copper line are 50 μm and 35 μm respectively. The interaction between the current and the electrical resistance as a function of temperature results in a small measurable 3ω signal, which is used to calculate the thermal conductivity of the zeolite membrane.

Figure 3 shows the thermal conductivities of the uncalcined and the calcined MFI membranes as a function of temperature. The conductivity of the unclacined MFI membrane, which

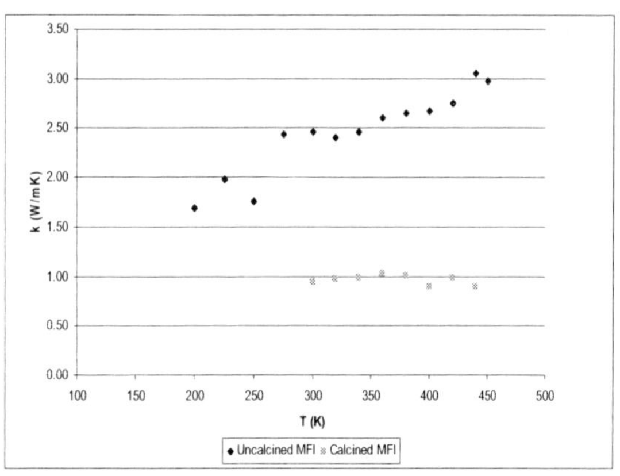

Fig. 3. Thermal conductivity of uncalcined (above) & calcined (below) MFI membrane as a function of temperature.

contains organic tetrapropylammonium (TPA) cations occluded within its pore spaces, is higher than that of the calcined MFI films which do not contain TPA. This is consistent with the generally observed increase in thermal conductivity with decreasing material porosity. Zeolites and other nanoporous materials present theoretical challenges in describing thermal transport, since they have a number of unusual characteristics (e.g. ordered nanoscale pores, negative thermal expansion) [4]. Our measured values are the highest reported thermal conductivity values for zeolite materials. One possible reason is that previous measurements were made by compacting powder samples into disks whose effective conductivity is measured and used to extract the intrinsic conductivity by a model that involves several assumptions regarding the structure of the disk and the thermal transport mechanisms [4,5]. However, the growth of oriented polycrystalline zeolite films leads to close-packed crystals with high preferred orientation. Additionally, the 3ω method can be used to extract both the in-plane and out-of-plane transport properties. We are currently working on quantitatively explaining the observed thermal conductivity behavior in terms of the phonon transport and scattering phenomena occurring in the crystals. We believe that the approach of combining thin film synthesis, the 3ω method, and theoretical analyses, will provide valuable insight into thermal transport in this industrially important class of materials with potential application in sorption-based heat pumping and refrigeration systems.

ACKNOWLEDGMENTS

We acknowledge financial support from the Georgia Institute of Technology and the National Science Foundation.

REFERENCES

[1] Davis, M. E. *Nature* **2002**, *417*, 813.

[2] Lai, Z. P.; Bonilla, G.; Diaz, I.; Nery, J. G.; Sujaoti, K.; Amat, M. A.; Kokkoli, E.; Terasaki, O.; Thompson, R. W.; Tsapatsis, M.; Vlachos, D. G. *Science* **2003**, *300*, 456.

[3] Hauer, A. *IES, ECES IA Annex 17, 3rd Workshop*, 1-2 October 2002, Tokyo, Japan.

[4] Murashov, V.; White, M. A. *Materials Chemistry and Physics* **2002**, *75*, 178

[5] Murashov, V. *J.Phys.:Condens. Matter* **1999**, *11*, 1261

[6] Nair, S.; Tsapatsis, M.; in *Handbook of Zeolite Science and Technology* (S. M. Auerbach et al, Eds.) Marcel Dekker: New York, 2003.

[7] Cahill, D. G. *Rev. Sci. Instrum.* **1990**, *61*, 802.

NANO2004-46097

Is there a critical size in nano grained metals for ductile to brittle transition?

Aman Haque and Taher Saif
Mechanical and Industrial Engineering
University of Illinois at Urbana-Champaign

Nanoscale metal films and electrodes are extensively used in today's micro and nano electronics as well as nano mechanical systems. These metal structures are usually polycrystalline in nature with nano scale grains connected to each other by grain boundaries. The small size offers large grain boundary to volume ratio that is likely to affect the metal properties significantly. Here, we discuss the role of grain size and boundaries in determining the mechanical behavior of metals, such as elasticity and yielding.

We first develop a novel micro apparatus that allows uniaxial tensile testing of free standing nano scale metal specimens as thin as 30nm. There are two unique features of the apparatus: (1) The specimen and the sensors to measure forces and displacements are co-fabricated – a paradigm shift in materials testing. Co-fabrication avoids the problem of handling the small sample, and suppresses any misalignment errors by five orders of magnitude [1], and (2) Testing can be carried out in-situ in SEM (Scanning Electron Microscope) or TEM (Transmission Electron Microscope). The latter allows visualization of the micro structure at sub-nanometer resolution while the stress-strain response is measured. Thus, the evolution of microstructure can be correlated with the stress-strain response to reveal the mechanism of deformation at nano scale. Figure 1 shows the apparatus with an aluminum sample, 100nm thick, 10μm wide and 250μm long.

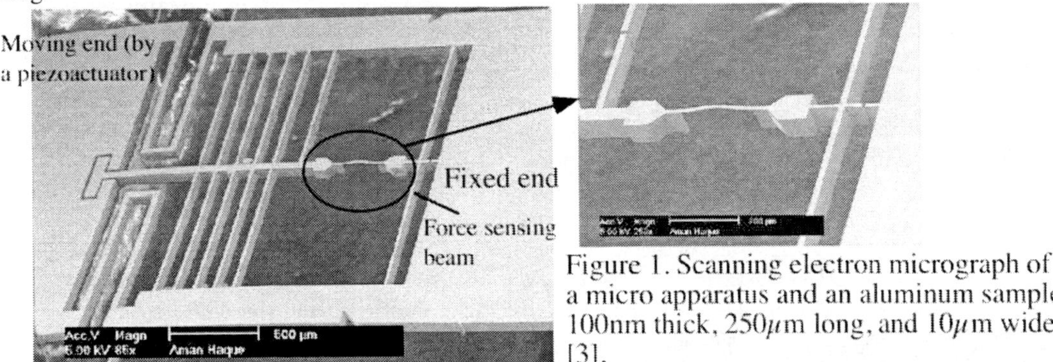

Figure 1. Scanning electron micrograph of a micro apparatus and an aluminum sample, 100nm thick, 250μm long, and 10μm wide [3].

Using the apparatus, we measure the stress-strain response of sputter deposited aluminum films with thickness 30-200nm with the corresponding grain size of 10-100nm. Figure 2 shows the result for 200nm and 50nm thick samples. Figure 3 shows the in-situ test result for 100nm thick sample with the TEM images. Note that there are no dislocations detected in the sample, even near the tip of the crack growing through the grains and grain boundaries. No void or crack is visible, nor any crack appearing during loading except prior to failure.

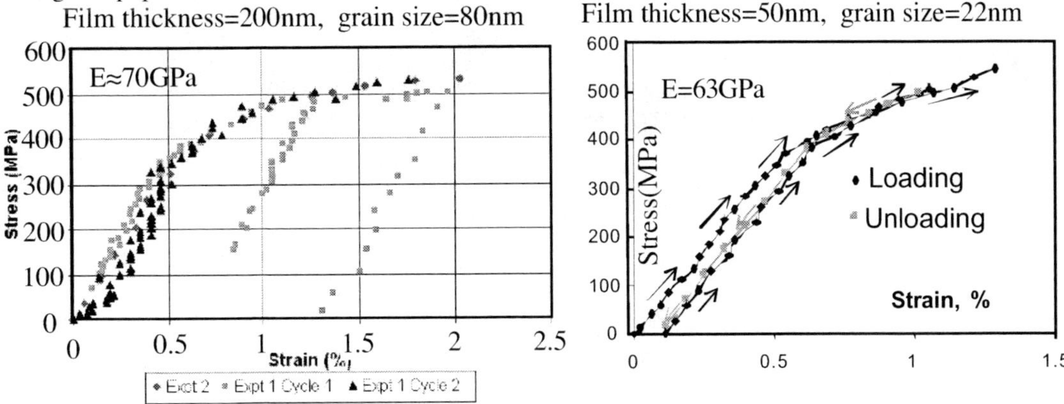

Figure 2. Stress-strain response of aluminum films. Small grained sample shows non-linear elastic behavior, in sharp contrast with larger grained (80nm) sample.

From a series of experiments, we find that as the grain size is decreased, (1) elastic modulus decreases by a small fraction, and (2) yield stress increases, reaches a maximum, and then decreases if yielding is defined as the deviation from the linear stress-strain behavior, and (3) metal behaves as no-linear elastic, i.e., it traces the loading stress-strain path when unloaded.

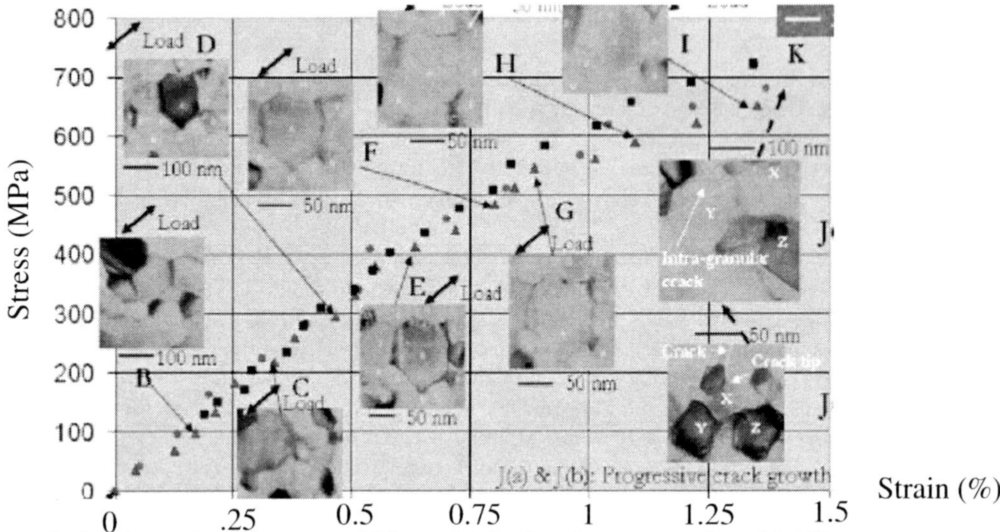

Figure 3. In-situ uniaxial test on 100nm thick aluminum sample with 50nm grains

As grain size decreases, it becomes energetically unfavorable to sustain and generate dislocations within grains. Hence dislocation slip ceases to act, and the metal begins to show lack of ductility. With decreasing grain size, grain boundaries become abundant. The atoms near the vicinity of the boundaries are at higher energy states due to lattice distortions. Thus, some atoms are further apart from the equilibrium distance, while some others are closer to each other to maintain equilibrium. Such atoms may span one or two atomic layers, but the abundance of grain boundaries render them significance in determining metal behavior. It can be shown that the higher energy state of the atoms render the boundary region more compliant. Thus, at small strains, metal shows lower elastic modulus. With increasing strain, grain boundaries become more compliant due to increased separation between the atoms, and metal shows softening non-linearity in the stress-strain response, all elastically. With unloading, the same stress-strain path is traced with little plastic deformation. Smaller the grains, higher is the abundance of grain boundaries, and softer the metal becomes. Thus, there is a critical size for the grains where the metal behavior transits from ductile to brittle non-linear elastic. As grain size is reduced from micro scale, a critical dimension is reached when the size is too small for dislocations to exist, but not small enough for grain boundary mechanisms to fully contribute to deformation. This is the critical grain size at which the metal is strongest [2,3]. Above this size, metal behaves as ductile, and below as non-linear elastic and brittle. Experimentally, it appears that for aluminum the critical size is around 50nm.

[1] M. A. Haque and M. T. A. Saif. (2002) *Expt. Mech.*, **42**, 123-128.
[2] V. Yamakov and D. Wolf and S. R. Phillpot. (2003) *Phil. Mag. Lett.*, **83**, 385-393.
[3] A. Haque, M. A. and M. T. A. Saif. (2004) *Proceedings of the National Academy of Science*, **101**, 6335-6340.

NANO2004-46008

NANOLITHOGRAPHY USING HIGH TRANSMISSION NANOSCALE RIDGE APERTURES

Sreemanth Uppuluri
School of Mechanical Engineering
Purdue University
West Lafayette

Eric X. Jin
School of Mechanical Engineering
Purdue University
West Lafayette

Xianfan Xu[*]
School of Mechanical Engineering
Purdue University
West Lafayette

In this paper, we describe using high transmission nanoscale apertures of C and H shapes for nanolithography applications. We demonstrate that these ridge apertures provide a highly localized and intense light spot that can be used in lithography experiments.

Near-field optical lithography comes under the category of scanning probe technologies proposed for overcoming the diffraction limit that is encountered in traditional optical lithography. Some of the near-field techniques involve illuminating a thin resist layer by light emitted either from the tip of a metal coated optical fiber [1] or from regular shaped apertures [2]. The main drawbacks of these approaches are that the process is a serial one and/or the transmission efficiency of the apertures is very low.

Numerical studies have predicted high transmission and field concentration capabilities of ridge apertures of C, H and bow-tie shapes [3]. Calculations showed that these sub-wavelength apertures have extraordinary field localization and transmission properties at visible wavelengths compared to ordinary square or circular shaped apertures. Thus such apertures allow sub-diffraction lithography to be performed at lower fluence levels of the incident laser. Also, patterning a mask with an array of such apertures allows parallel operation.

Figure 1 shows the geometry of C and H apertures used in our experiments. A contact lithography scheme is followed by using Cr-coated quartz wafers as masks and resist coated wafers as substrates. We employed FDTD calculations to determine the sizes of the C and H apertures for obtaining high transmission and field localization at a wavelength of 457 nm and patterned these apertures in the Cr-film by FIB milling. The performance of the C and H apertures are juxtaposed with that of sub-micron size square apertures of 50 nm size and large rectangular apertures (12 μm × 4 μm).

The results of the experiments shown in Tables 1 and 2 correspond to the polarization in the y- and x-directions of the incident laser beam (the directions are the same as shown in Fig. 1 with respect to the mask). During the experiments, the laser intensity is maintained constant and the exposure durations are varied. The experiment results show that the C and H apertures exhibit high transmission and localization properties only when exposed with light polarized in the y-direction. This behavior has been proved by simulations as well [3].

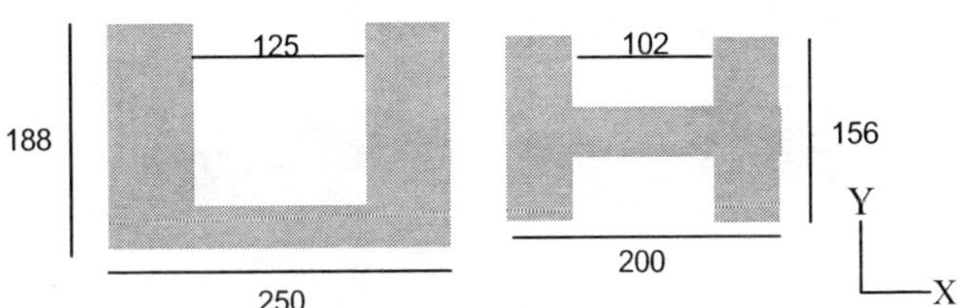

Figure 1: Dimensions of the C and H Apertures in nm

Best lithography results for C and H apertures are obtained for exposure times around 15 seconds. At these doses, the rectangular apertures developed partially, thus indicating that the doses are around the threshold value required for complete development. On the other hand the sub-micron square apertures did not yield any holes even when exposed for 120 seconds for either polarization. *This proves the high transmission of the C and H apertures*. Figures 2(a) and 2(b) show the holes in the resist corresponding to the C and H apertures for an exposure time of 15 seconds. The sizes of the holes are 95 nm × 145 nm and 85 nm × 135 nm corresponding to C and H apertures respectively. *This proves the field localization capabilities of the apertures*.

[*] Author to whom correspondence must be addressed, Phone: 765-494-5639, email xxu@ecn.purdue.edu

Table 1: Exposure results for y-polarization of the incident laser light.

C-aperture | H-aperture

No result	Irregularly Developed Holes with shallow depth	Size = 85 nm × 135 nm × 10 nm	Size = 107 nm × 240 nm × 57 nm	Size = 167 nm × 315 nm × 94 nm
No result	Irregularly Developed Holes with shallow depth	Size = 95 nm ×145 nm × 28 nm	Size = 178 nm × 237 nm × 68 nm	Size = 159 nm × 315 nm × 110 nm

0	5	10	15	20	30

Exposure time in seconds

Table 2: Exposure results for x-polarization of the incident laser light

C-aperture | H-aperture

No result	No result	No result	No result	No result
No result	No result	No result	No result	No result

0	5	10	15	30	120

Exposure time in seconds

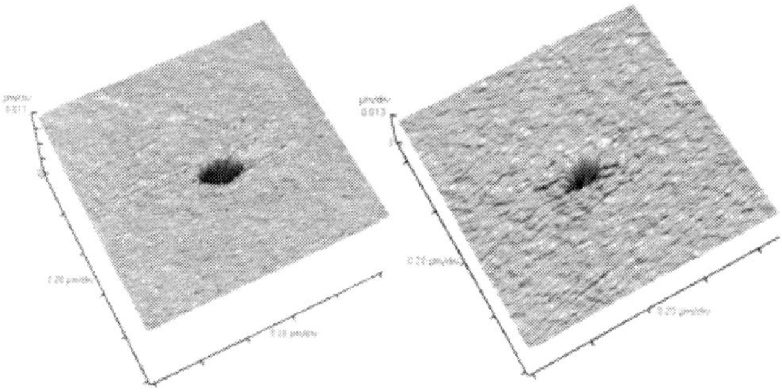

Figure 2: AFM images of holes in the resist corresponding to C and H apertures for exposure duration of 15 seconds

FDTD calculations are carried out by using the ridge aperture geometries displayed in Fig. 1. A semi-infinite photoresist layer is assumed to be placed at 10 nm below the apertures (Although it is contact lithography, there will still be some gap between the aperture and photoresist due to surface roughness). The dielectric constant of photoresist is chosen to be 2.82 according to the specification of S-1805. The FDTD simulations follow the procedures described previously [3]. The experimental results could be qualitatively compared with FDTD computations. Plots in Fig. 3 show the amplitude distributions of y and z components of E field at 10 nm below (at the resist surface) the H aperture. Similar near field can also be found in C aperture (not shown here) It can be seen that

the y-component of the electric field is well confined below the gap region defined by the ridges of H aperture, which gives a hot spot much less than the size of the H aperture itself. The elongation of the hole in the y-direction found in the resist could be explained as a result of the spreading of the z component of the transmitted field (Fig. 3(b)) along the bottom of the ridges (The pointing vector for Ez component is in y-direction). Due to the same reason, the hole produced by C aperture is also elongated in y-direction as seen in the Fig. 2(a).

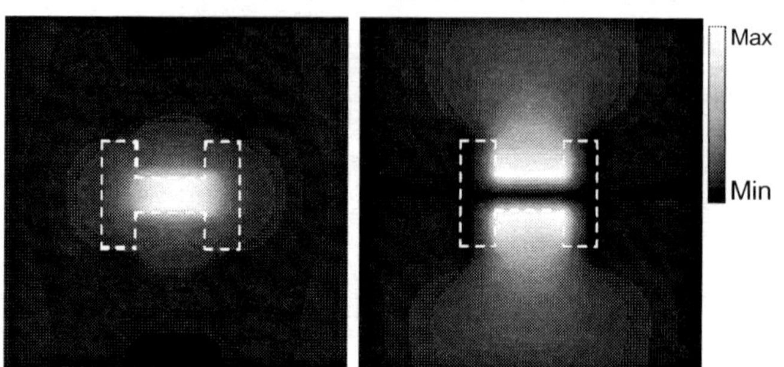

Figure 3: FDTD computations showing the distributions of the y and z components of the electric field for H aperture with presence of photoresist 10 nm underneath.

In summary, this work presents the first demonstration of field concentration and transmission enhancement of C and H apertures in a contact lithography experiment. The C and H apertures yielded holes of about 100 nm in dimension. Under identical exposure conditions, sub-micron square holes of similar dimensions did not yield any result. The experiment thus demonstrated field concentration and enhancement of C and H apertures.

ACKNOWLEDGMENTS: Support to this work by the National Science Foundation is acknowledged.

REFERENCES:
1. Rudman, M., Lewis, A., Mallul, A., Haviv, V., Turovets, I., Shchemelinin, A., Nebenzahl, I., 1992, "Near-field subwavelength micropattern generation: Pipette guided argon fluoride excimer laser microfabrication", J. of Applied Physics, **72**, pp. 4379-4383.
2. Blaikie, R. J., Alkaisi, M. M., McNab, S. J., Cumming, D. R. S., Cheung, R., Hasko, D. G., 1999, "Nanolithography Using Optical Contact Exposure in the Evanescent Near Field", Microelectronic Engineering, **46**, pp 85-88.
3. Jin, E. X., Xu, X., 2004, "Finite-Difference Time-Domain Studies on Optical Transmission through Planar Nano-Apertures in a Metal Film", Jpn. J. of Applied Physics, **43**, pp. 407-417.

NANO2004-46021

LARGE-SCALE ASSEMBLY OF CARBON NANOTUBES

T. El-Aguizy
Massachusetts Institute of Technology
Department of Mechanical Engineering
Cambridge, MA 02139

Sang-Gook Kim
Massachusetts Institute of Technology
Department of Mechanical Engineering
Cambridge, MA 02139
Email:sangkim@mit.edu

ABSTRACT

The scale decomposition of a multi-scale system into small-scale order domains will reduce the complexity of the system and will subsequently ensure a success in nanomanufacturing. A novel method of assembling individual carbon nanotube has been developed based on the concept of scale decomposition. Current technologies for organized growth of carbon nanotubes are limited to very small-scale order. The nanopelleting concept is to overcome this limitation by embedding carbon nanotubes into micro-scale pellets that enable large-scale assembly as required. Manufacturing processes have been developed to produce nanopellets, which are then transplanted to locations where the functionalization of carbon nanotubes are required.

Keywords: Carbon nanotubes, Directed assembly, Multi-scale systems integration, Nanopelleting

DESIGN

Design and manufacture of nanostructure require much more precision than that of macro and microstructures. Many research groups have tried to build and control nanostructures bottom-up in this regard. The quantum dots, nanoparticles, nanotubes and nanowires are made with sub-nanometer accuracy, but have limited repeatability only within a short-range order. In order to benefit from the nanotechnology, nanostructures need to be integrated to the micro/macro-scale platforms, where the top-down and bottom-up processes meet.

Carbon nanotubes have been reported to have many unique and useful properties. Their integration into real applications requires robust large-scale assembly techniques. A key challenge to manufacture products with CNTs is handling of nano-scale objects during assembly. In particular, the construction of regular arrays of uniformly aligned CNTs requires control over the CNT orientation and location. Current techniques for large-scale CNT assembly utilize guided liquid dispersions, gas flow and electric field that do not provide deterministic control over the CNTs. [1-3] Other methods for CNT assembly control orientations during the growth step, particularly using plasma-enhanced CVD systems with pre-patterned catalysts. [4,5] Such techniques are effective in creating aligned CNTs with accurate location control, but are limited in their scale and order to very small areas. We developed a technique, termed *nanopelleting*, for the integration of CNTs into micro-scale pellets that can be transplanted and deterministically assembled.

The *nanopelleting* concept provides specific control over the length, alignment, and position of assembled CNTs. A nanopellet consists of a CNT embedded within a micro-scale pellet. The pellet serves as a carrier for the CNTs when they are integrated into multi-scale systems. Further, by decoupling the growth of CNTs from their use, pellets can be harvested and reassembled in bulk, enabling the fabrication of devices with long-range order and large scale.

PROCESSES and RESULTS

The fabrication process developed for nanopelleting relies on a combination of standard microfabrication steps as shown in Fig. 1. The steps to create a nanopellet include the creation of a trench within which CNTs are grown via PE-CVD process. The trenches are then filled with a spin coated epoxy polymer (M-Bond 610), encapsulating the CNTs. The surface is then planarized via CMP process to create pellets cast to the geometry of the trenches and CNTs with uniform alignment and length. The pellets and the embedded CNTs can then be released using a XeF_2 etch to selectively undercut only the silicon. Similarly, the pellet body material can be selectively removed without damaging the CNTs.

The results of this fabrication sequence demonstrate the capability of the process to encapsulate CNTs within the pellets that are then planarized to a uniform length, as shown in Fig. 2 (c). The pellet is removed from the original substrate by selectively under-etching the silicon. After the release, the pellets are then ready to be harvested and transplanted to an acceptor substrate. Variations in this process are also considered, including the use of additive techniques without trenches.

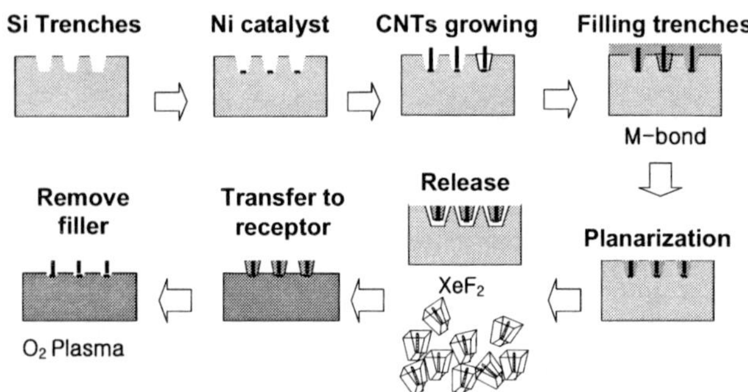

Figure 1: Representative process flow for creation of Nanopellets

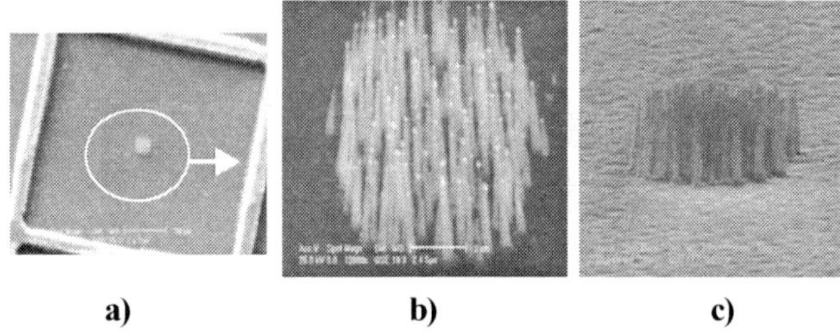

a)   b)   c)

Figure 2: a) A silicon trench with CNTs grown as a bundle at the center. Side view of CNTs before, (b), and after, (c), planarization showing uniform CNT length.

For transplanting of the pellets, we mechanically manipulated the micro-scale pellets onto an acceptor substrate with a pre-etched 'T' shape, as shown in Fig. 3 (a). Figure 3 (b) shows schematic view of transplanting a nanopellet onto a template. A magnified view of the pellet and embedded CNTs is shown in Fig. 4 (a) and (b) before and after the removal of dielectric material. Pellets are therefore effective in acting as a carrier that allows deterministic assembly of CNTs. Coupled with many self-assembly methods for micro-scale structures [6], large-scale assembly of CNTs would become possible by the *nanopelleting* process.

With the concept of nanopelleting, we have shown the possibility of guided assembly of individual nanotube into a long-range ordered 2-D array. The results support the nanopelleting concept as a means to produce large-scale arrays of CNTs with long-range order and deterministic positioning and alignment. Such a technique would enable applications that require large-scale arrays of CNTs with long-range order. This includes field emission systems for displays, massively parallel nanoprobing systems, or others. Moreover, given the broader applicability of the nanopelleting concept to nanostructures, the technique could also be used for handling and fabrication of other nanostructures. Further work on refining this technique

for integration into actual applications is necessary, including high aspect ratio nanopellets.

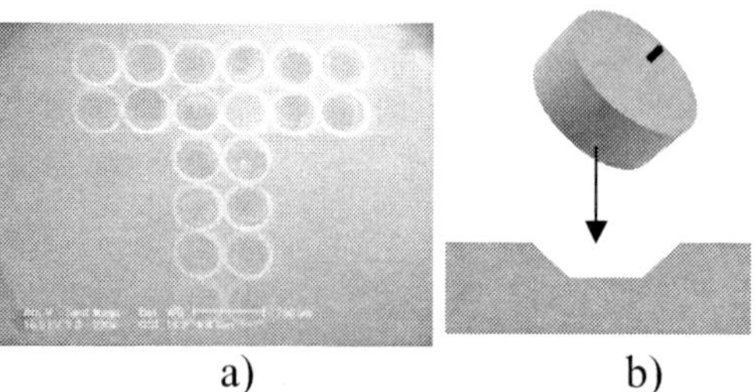

a)   b)

Figure 3: Transplanted pellet arranged into receptor trenches forming 'T' shape

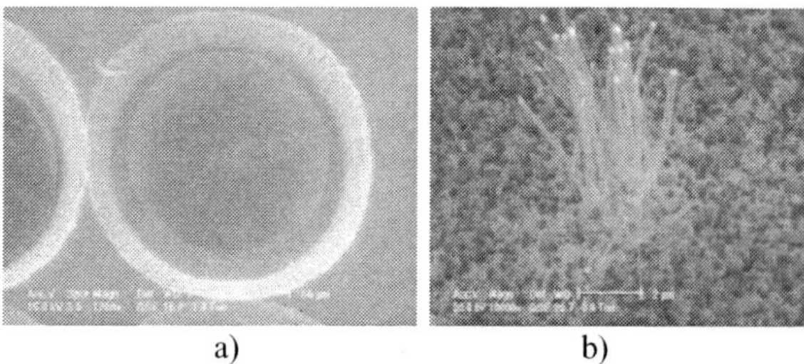

a)   b)

Figure 4: Pellet with CNTs in center, before XeF₂ etching, (a) and after XeF₂ etching, (b).

ACKNOWLEDGMENTS

This work is funded by the Deshpande Center at MIT and Center for Nanoscale Mechatronics and Manufacturing at KIMM. The contributions by Dr. Y.B. Jeon and J. H. Jeong are deeply appreciated.

REFERENCES

[1] Huang, S., Dai, L., Mau, A., 2002, "Controlled fabrication of aligned carbon nanotube patterns," Physica B, **323**, pp. 333-335.
[2] Chung J., Lee K., Lee J. and Ruoff R, 2004, "Toward Large-scale integration of carbon nanotubes," Langmuir, **20**, pp. 3011
[3] S. Rao, et al., "Large-scale assembly of carbon nanotubes", Nature, **425**, brief communications, pp. 36-37
[4] Ren, Z.F., Huang, Z.P., Xu, J.W., Wang, J.H., Bush, P., Siegal, M.P., Provencio, P.N., 1998, "Synthesis of Large Arrays of Well-Aligned Carbon Nanotubes on Glass," Science, **282**, pp. 1105-1107.
[5] Ren, Z.F., Huang, Z.P., Wang, D.Z., Wen, J.G., Xu, J.W., Wang, J.H., Calvet, L.E., Chen, J., Klemic, J.F., Reed, M.A., 1999, "Growth of a single freestanding multiwall carbon nanotubes on each nanonickel dot," Applied Physics Letters, **75**, pp. 1086-1088.
[6] Srinivasan, U., Liepmann, D., Howe, R.T., 2001, "Microstructure to substrate self-assembly using capillary forces," J. of Microelectromechanical Systems, **10**, pp. 17-24

NANO2004-46023

PLASMONIC LITHOGRAPHY

W. Srituravanich, N. Fang, C. Sun, S. Durant, M. Ambati and X. Zhang

NSF Center for Scalable and Integrated Nanomanufacturing, University of California, Los Angeles,
California 90095, http://microlab.seas.ucla.edu

ABSTRACT

As the next-generation technology moves below 100 nm mark, the need arises for a capability of manipulation and positioning of light on the scale of tens of nanometers. Plasmonic optics opens the door to operate beyond the diffraction limit by placing a sub-wavelength aperture in an opaque metal sheet. Recent experimental works [1] demonstrated that a giant transmission efficiency (>15%) can be achieved by exciting the surface plasmons with artificially displaced arrays of sub-wavelength holes. Moreover the effectively short modal wavelength of surface plasmons opens up the possibility to overcome the diffraction limit in the near-field lithography. This shows promise in a revolutionary high throughput and high density optical lithography.

In this paper, we demonstrate the feasibility of near-field nanolithography by exciting surface plasmon on nanostructures perforated on metal film. Plasmonic masks of hole arrays and "bull's eye" structures (single hole surrounded by concentric ring grating) [2] are fabricated using Focused Ion Beam (FIB). A special index matching spacer layer is then deposited onto the masks to ensure high transmissivity. Consequently, an I-line negative photoresist is spun on the top of spacer layer in order to obtain the exposure results. A FDTD simulation study has been conducted to predict the near field profile [3] of the designed plasmonic masks. Our preliminary exposure test using these hole-array masks demonstrated 170 nm period dot array patterns, well beyond the resolution limit of conventional lithography using near-UV wavelength. Furthermore, the exposure result obtained from the bull's eye structures indicated the characteristics of periodicity and polarization dependence, which confirmed the contribution of surface plasmons.

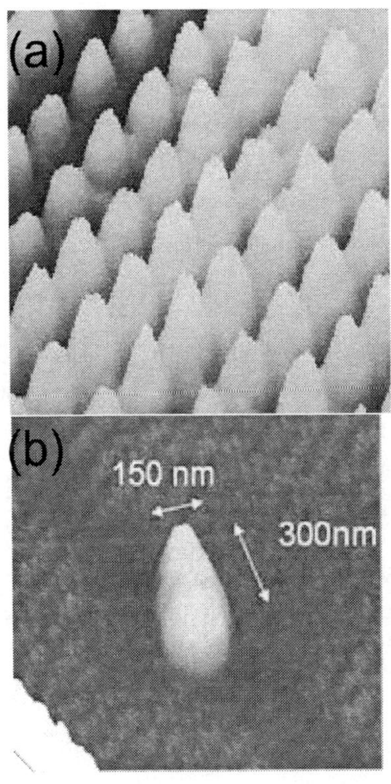

Fig 1. (a) AFM image of the patterned dot array with 170 nm period. (b) AFM image of exposure result from bull's eye structure.

ACKNOWLEDGMENTS

The authors are grateful for Fujino Makoto for the valuable help and inspiring discussions. This work was supported by MURI (Grant # N00014-01-1-0803), NSF on Nanoscale Science and Engineering Center (NSEC) (Grant # DMI-0327077), and NSF (Grant # DMI-0218273).

REFERENCES

1. Ebbesen, T.W.; Lezec, H. J.; Ghaemi, H. F.; et al, **"Extraordinary optical transmission through sub-wavelength hole arrays"**, Nature, 391(1998)667.
2. Thio T., Pellerin K.M., Linke R.A., Lezec H.J., Ebbesen T.W., **"Enhanced light transmission through a single subwavelength aperture."** Opt. Lett., 26(2001)1972.
3. Salomon L, Grillot F, Zayats AV, de Fornel F., **"Near-field distribution of optical transmission of periodic subwavelength holes in a metal film,"** Phys. Rev. Lett., 86(2001)1110.
4. Werayut Srituravanich, Nicholas Fang, Cheng Sun, Qi Luo, and Xiang Zhang. **Plasmonic Nanolithography**, Nano. Lett., *4* (2004)1085.

NANO2004-46029

Hypersonic Plasma Particle Deposition of Nanostructured Coatings and Micropatterns

J. Hafiz[1], R. Mukherjee[1], X. Wang[1], W. Mook[2],
J. V. R. Heberlein[1], P. H. McMurry[1], W.W. Gerberich[2] and S. L. Girshick[1]
[1]Dept. of Mechanical Engineering
[2]Dept. of Chemical Engineering and Materials Science
University of Minnesota, Minneapolis, MN, USA

Nanostructured materials are attractive candidates for advanced friction and wear-resistant coatings due to their potentially enhanced mechanical properties. We have developed a one-step method called hypersonic plasma particle deposition [1] to produce and deposit nanoparticles using a thermal plasma reactor. Particle synthesis is achieved by dissociating vapor phase reactants in the plasma, and quenching the hot gas in a supersonic nozzle expansion. Particles are deposited on a substrate by hypersonic impaction to form a film, or are deposited as micropatterns using a process called focused particle beam deposition [2]. In-situ particle size distribution measurements are performed using a sampling probe interfaced to an extraction/dilution system for measurement by a scanning electrical mobility spectrometer.

We have synthesized Si-Ti-N thin films of 20-50 μm thickness on Mo substrates at rates of 2-10 μm/min, depending on reactant flow rates, at substrate temperatures ranging from 250-850°C. In-situ particle size distribution measurements showed that particle size distributions peak around 10 nm under typical operating conditions (Fig. 1). Microstructural characterization of the films was performed using scanning electron microscopy (SEM), Rutherford backscattering (RBS), X-ray photoelectron spectroscopy (XPS) and X-ray diffraction (XRD). SEM results confirm the presence of nanostructured grains. XRD results show the presence of crystalline TiN and $TiSi_2$, while XPS measurements illustrate the presence of amorphous Si_3N_4. Film post-processing was carried out using in-situ plasma sintering and compression sintering. The effects of post-processing on film properties are being explored. Hardness of as-deposited films was evaluated by nanoindentation of polished film cross-sections. Measured hardness values, averaged over 10-15 locations for each film, equaled 22-26 GPa.

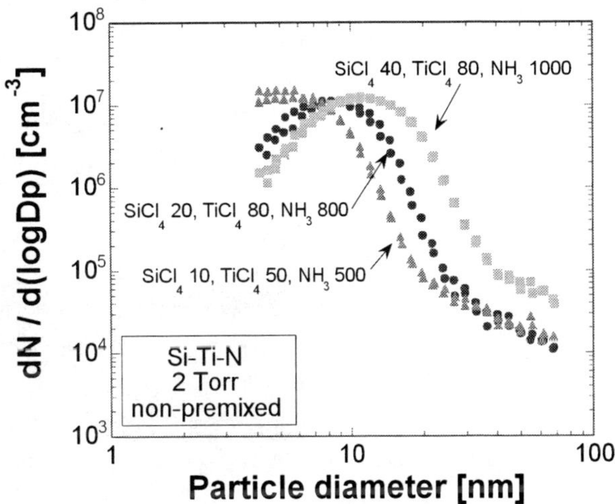

Figure 1. Measured size distributions of Si-Ti-N particles, sampled in-situ. Flow rates indicated are in sccm.

In the focused particle beam deposition process, the nanoparticles from the supersonic nozzle are collimated into a tight beam (~40 μm wide) using aerodynamic lenses. The focused nanoparticles are then deposited onto a motion-controlled substrate (typically a silicon wafer or TEM grids) to form micropatterns, with feature sizes down to few tens of microns. The materials that have been deposited using this process include silicon, titanium, their carbides and nitrides, and also their composites. We are currently pursuing micromolding of MEMS devices by filling micromachined silicon molds with focused nanoparticles. Fig. 2(a) shows a silicon gear micromold fabricated using standard microfabrication techniques, while Fig. 2(b) illustrates the cross section of a gear filled with Si-C nanoparticles using the focused nanoparticle beam. The silicon substrate was at room temperature during deposition.

In additional experiments, we also filled gear molds by replacing the molybdenum substrate used in the film deposition setup. This was done to utilize the high rate deposition capabilities and higher substrate temperatures attainable in the hypersonic impaction stage. Preliminary results (Fig. 3) illustrate a cross section image of a gear partially filled with a film of Si-Ti-N nanoparticles.

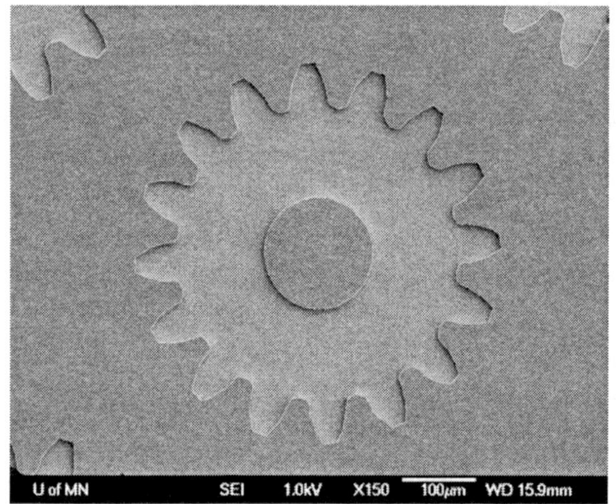

Figure 2(a). Micromachined silicon gear mold.

Figure 2(b). Cross-section of mold partially filled by thin layer of SiC nanoparticles from focused beam.

Figure 3. Cross-section of Si-Ti-N nanoparticle film deposited in "valley" of gear mold in high-rate deposition stage. Substrate temperature ~500-650°C, 1-minute deposition.

Partial support from NSF Grant No. DMI-0103169 is gratefully acknowledged.

[1] Rao, N.P., N. Tymiak, J. Blum, A. Neuman, H. J. Lee, S. L. Girshick, P. H. McMurry and J. Heberlein, 1998, "Experimental study of nanostructured silicon carbide film formation by hypersonic plasma particle deposition," Journal of Aerosol Science, **29**, pp. 707-720.
[2] Di Fonzo, F., A. Gidwani, M. H. Fan, D. Neumann, D. I. Iordanoglou, J. V. R. Heberlein, P. H. McMurry, S. L. Girshick, N. Tymiak, W. W. Gerberich and N. P. Rao, 2000, "Focused nanoparticle-beam deposition of patterned microstructures," Applied Physics Letters, **77**, pp. 910-912.

NANO2004-46040

Performance of a Dual Side Substrate Metrology System for Micromachining Lithography

Daniel Schurz, Warren W. Flack
Ultratech, Inc.
San Jose, CA 95134

Advances in micromachining (MEMS) applications such as optical components, inertial and pressure sensors, fluidic pumps and radio frequency (RF) devices are driving lithographic requirements for tighter registration, improved pattern resolution and improved process control on both sides of the substrate. Consequently, there is a similar increase in demand for advanced metrology tools capable of measuring the Dual Side Alignment (DSA) performance of the lithography systems.

There are a number of requirements for an advanced DSA metrology tool. First, the system should be capable of measuring points over the entire area of the wafer rather than a narrow area near the lithography alignment targets. Secondly, the system should be capable of measuring a variety of different substrate types and thicknesses. Finally, it should be able to measure substrates containing opaque deposited films such as metals.

In this paper, the operation and performance of a new DSA metrology tool is discussed. The UltraMet 100 offers DSA registration measurement at greater than 90% of a wafer's surface area, providing a true picture of a lithography tool's alignment performance and registration yield across the wafer. The system architecture is discussed including the use of top and bottom cameras and the pattern recognition system. Experimental data is shown for tool performance in terms of repeatability and reproducibility over time. The requirements for tool accuracy and methods to establish accuracy to a NIST traceable standard are also discussed.

Key Words: Dual Side Alignment, DSA, repeatability, reproducibility, accuracy, artifact

1.0 INTRODUCTION

1.1 Dual Side Alignment

In the expanding MEMS and nanotechnology markets, a number of device types require alignment and exposure between the front and back surfaces of a substrate. For example, DSA is used in creating system-on-chips in three-dimensional integrated circuits for successive bonding of silicon-on-insulator (SOI) device layers [1] and for Radio Frequency devices to minimize attenuation by metal deposition [2]. DSA is also used for ink jet heads to align backside channels for ink flow to front-side nozzles [3]. Other devices requiring dual-side alignment are inertial accelerometers, pressure sensors, variable optical attenuators (VOAs), magnetic and optical-networking components and microshutter arrays [2].

Until recently, the resolution requirements and alignment tolerances for DSA have not been critical, therefore, metrology to verify alignment has not been widely utilized. Now, with shrinking geometries, DSA alignment is becoming more

103

stringent and metrology becoming more important. Figure 1 shows an Ultratech customer survey of predicted DSA requirements. The maximum and minimum curves represent the range of customer DSA requirements over time. A requirement for 1.0 μm occurs as early as 2006. As the resolution and alignment requirements have become more challenging, a number of companies are moving to projection lithography rather than full wafer aligners for volume production. Sample inspection is becoming more widespread, providing the feedback needed to implement statistical process control (SPC) in DSA process steps. In order to determine the linear error components (scaling, rotation, orthogonality and mean), the alignment should be checked at a minimum of six points across a wafer [4]. This is especially critical for warped or bonded wafers, since two point metrology at the edges of the wafer cannot give accurate or complete alignment information. And, as companies look at continuing improvement in yield, linear overlay errors will need to be more stringently evaluated.

Since advanced MEMS devices are not yet in high production volume, automated DSA metrology is not a requirement for most manufacturers. A lower priced manual metrology tool that can be upgraded in the future is sufficient. However, the tool needs to allow multiple points to be measured across the wafer. It is also necessary for the data to be analyzed automatically. It is anticipated that in the future, a fully automated metrology tool will be required.

This paper will discuss the design and performance of a DSA metrology tool. It will provide measurement data showing system performance in terms of alignment repeatability and reproducibility for multiple sites across the wafer.

1.2 Performance Requirements

The performance of any metrology tool is based on the ability to measure a feature (or its relative position) with repeatable accuracy under two conditions. The tool must be able to make repeated measurements within a specified tolerance relative to the mean, with no adjustments to the tool or movement of the device being measured (repeatability). It must also be able to measure within a given tolerance over time with the measurement feature removed and reloaded onto the tool (reproducibility). The results obtained in reproducibility encompass all contributing errors inherent in the tool and in the measurement process, including operator error caused by substrate reloading [5,6]. Because a tool's precision (P) is the total measurement variation, it can then be calculated as [6]:

$$P = \sqrt{(\sigma_{repeat})^2 + (\sigma_{reprod})^2} \qquad (1)$$

where σ_{repeat} is the sigma of repeatability and σ_{reprod} is the sigma of reproducibility.

Tool accuracy is a combination of random and systematic sources of error compared to an established reference. At this time, there are no widely accepted standards available that provide referenced distances between points on opposite surfaces of a thin substrate. One method to achieve accuracy is by careful calibration of each imaging subsystem as described in section 2.1. Another is to create an artifact that provides a fixed distance between two features that can be measured both on an industry accepted X/Y pattern placement tool calibrated to a NIST standard, and on the metrology tool studied. Both methods are discussed in this paper.

The first generation of the UltraMet system is designed to measure alignment where process tolerances are 1.0 μm or greater. Accordingly, to achieve a precision to tolerance ratio less than 30%, repeatability and reproducibility for this tool are specified at 0.10 μm 3σ and 0.30 μm 3σ respectively. The precision to tolerance ratio is calculated as follows:

$$\frac{P}{T} = \frac{100 \times 6\sigma_{MS}}{T} \qquad (2)$$

where σ_{MS} is the precision of the measurement system.

1.3 Tool Description

The UltraMet 100 DSA tool provides the capability to measure the alignment between patterns printed on the top and bottom of a substrate as shown in Figure 2. The tool's basic configuration is shown in Figure 3. The X-Y substrate stage is located on a granite block for stability. The metrology is performed by two optical microscopes with 12.5X magnification. One views the top of the substrate and is movable in the Z-axis, while a second microscope simultaneously views the bottom of the substrate and is stationary in X, Y, and Z. Because two microscopes are used, the tool is capable of measuring a wide variety of opaque and transparent substrates and film types.

The substrate holder is a glass material that is transparent at the operating wavelength of the microscopes. This allows measurement of the bottom surface of a substrate at nearly any site within the six inch holder diameter. It is easily reconfigured for substrate sizes ranging from 2 inches to 6 inches, round or square, and is capable of handling substrate thickness up to 2 mm. A substrate can be rotated to 0° and 180° orientations for Tool Induced Shift (TIS) correction [7].

The computer system for controlling and operating the tool includes a video frame grabbing board for capturing images from the Charge Coupled Device (CCD) cameras using Cognex's Patmax$^{\circledR}$ image processing software. Patmax identifies key features in an object and correlates their spatial relationships to match a trained image to the image on the substrate, and is unaffected by the objects angle, size or appearance [8]. Patmax's tolerance of changes in an object's orientation allows for rapid target acquisition. The computer is located on a separate cart from the stage and measurement units. The computer system provides a monitor for visual display of the video image from both top and bottom cameras simultaneously.

The pattern recognition system, or Machine Vision System (MVS), provides a high degree of precision and versatility in registration measurement using a wide range of target shapes and sizes. Nearly any unique feature within an approximate 300 µm viewing window, ranging from 20 to 100 µm in size and with line-width greater than 4.0 µm can be successfully acquired by the MVS, including existing metrology or device features within any die on the substrate [9]. The top and bottom CCD cameras can be trained to acquire different target shapes simultaneously which provides more flexibility in job setup. A reference fiducial consisting of a grid of measurement structures for CCD pixel calibration is embedded in a peripheral area of the stage.

2.0 EXPERIMENTAL METHODS

2.1 Ultramet 100 Operation

2.1.1 Calibration:

After initial system setup, there is a single calibration routine required on daily or weekly intervals, depending on the tool usage. Pixel calibration is needed to translate pixel space (camera coordinates) that the computer uses to calculate physical space (microns). The single sided calibration process "learns" the mapping from the camera's pixel plane to the image plane on the wafer surface. It is used to account for subtle differences in optical magnification, position, rotation, and camera tilt between the top camera and the bottom camera.

Pixel calibration is performed using an array of reference fiducials spaced 100 µm apart embedded in a peripheral area of the substrate stage. Two images are taken simultaneously, one with the top camera and one with the bottom camera looking at the same array of fiducials. The bottom camera is used as the base point of reference (the exact center of the image is considered the XY plane 0,0 point). The bottom image is compared to the top image, looking for differences in translation, rotation, tilt, and spacing between fiducials. A Cognex software algorithm is utilized to calculate and store these differences. Calibration is complete and any further measurements will have the stored algorithm applied [10]. On screen instructions guide the operator through the calibration procedure, which takes less than 5 minutes.

2.1.2 Measurement:

Job files are created by the operator that specify the desired number of measurements to be repeated at a single site. Any number of repeated measurements can be set depending on the user's requirements for a statistical analysis confidence level [11]. The system software provides a text file containing each raw measurement point at each site, as well as a mean value and 3σ value for the site. The operator also selects a "step list" of the number and order of sites to be measured. Step lists are written as text files and placed in a directory for attachment to a job file. The system software uses the step list to guide the operator through the measurement routine. The system supports importing a bitmap file that depicts the sites to be measured from any commercial graphic software program. The bitmap file can be attached to the job file and used to guide the operator.

The operator has the option of using a target feature from an existing target library, or training on a new device or metrology feature on the substrate. The training process creates an image file that is placed in a directory and then attached to the job file.

Because the UltraMet 100 uses two microscopes to measure offsets between targets located on both sides of a substrate, a Tool Induced Shift (TIS) calculation must be made and included in the final offset calculation. When the completed job file is selected for the measurement routine, a substrate is loaded on the tool in a 180° rotated orientation, moved to the first target site, focused and measured. The substrate is then removed and reloaded at a 0° orientation, moved to the original site and re-measured. Then the substrate is moved to the remaining designated sites and measured. X and Y offsets and rotational differences from the first site measured at 0° and 180° are used to calculate TIS correction offsets that are applied to all measurement sites by the tool software at the end of the measurement routine. The TIS measurements add approximately one minute to the measurement routine for each substrate.

2.1.3 Data Analysis:

An ASCII measurement file containing all raw measurement results, TIS correction values and all TIS adjusted measurements for each site can then be viewed on-screen. Values are calculated in both the X and Y-axes. The data files can be loaded into any spreadsheet or statistical software package for statistical analysis.

2.1.4 Accuracy:

Because no industry standard exists that allows for simultaneous measurement of the offset between features on two sides of a substrate, an artifact was designed consisting of a chrome on quartz reference wafer with pairs of target features that could be measured on a Leica LMS2020 X/Y pattern placement metrology tool. The LMS2020 is specified for long term repeatability at 12 nm, and stage position accuracy at 25 nm. It is calibrated to a NIST traceable standard. The chrome on quartz wafer was manufactured with two different target feature shapes separated by 40 µm in the X axis and a zero Y axis offset. The wafer was then measured on the UltraMet 100 using the previously discussed TIS corrections. Both clear and chrome pairs of target features were evaluated to determine system sensitivity to target polarity.

2.2 Wafer Processing

The test wafers were SEMI-Standard, dual-side polished, 6 inch silicon wafers. These substrates are 650-700 µm thick with 1500 Å of thermal oxide and a TIR/TTV < 1 µm. They were patterned on both top and bottom surfaces using an Ultratech NanoTech 160 Wafer Stepper[®]. The optical specifications for the NanoTech 160 are shown in Table 1. The NanoTech 160 uses broadband gh-line illumination with DSA alignment capability for backside exposure. The photoresist for both top and bottom lithography was Shipley S1808 and the wafers were coated with 1.2 µm thick on both sides using the process and equipment described in Table 2. The resist thickness was measured on a Nanometric 8100X.

2.3 Artifact Processing

The chrome on quartz artifact was manufactured by first generating 10X reticles on a TRE 220 Mask Pattern Generator. The reticles were imaged onto the chrome wafer using an ASET 645 Step & Repeat camera. The wafer was coated with 5300 Angstroms of AZ1518 photoresist and developed in Microchrome's PPD455 developer. The chrome was etched for 30 seconds at ambient temperature in Micorchrome CEP200 etchant.

2.4 Experimental Data

Repeatability was calculated at two sites on a test wafer which were each measured thirty successive times with no movement between measurements. TIS measurements were made at 0° and 180° orientations as described in section 2.1.2. The wafer was removed and reloaded (cycled) and the entire measurement procedure repeated 30 times. After TIS correction of the 30 successive site measurements, a 3σ value was calculated for each site in both X and Y-axes. The highest 3σ values were recorded.

Reproducibility was calculated using the same TIS corrected measurements used for repeatability. A 3σ value was calculated for the entire set of 30 measurements times 30 cycles at one site on a test wafer.

Accuracy was determined using one pair of target features on the chrome on quartz artifact to compare the measurements taken on the LMS2020 and the UltraMet 100. The LMS2020 measurement value is the average value of 20 scans taken at a single site. The repeatability of the LMS2020 was confirmed by measuring the same site 20 times over the course of 4 days. The UltraMet 100 measurement value is the average of 30 repeated measurements taken at a single site with TIS correction values applied.

3.0 RESULTS AND DISCUSSIONS

3.1 Repeatability

Tool repeatability can be seen in the box and whisker plots in Figures 4a and 4b. These plots represent measurement data from a single wafer measured through 30 cycles on a single tool as described in section 1.2. Each cycle consists of 30 repeatability measurements taken at a site without any wafer movement. The top, bottom and middle line of each box corresponds to the 75th percentile (top quartile) 25th percentile (bottom quartile) and 50th percentile (median) of the 30 measurements [11]. The whiskers extend from the 10th percentile (bottom) and the 90th (top). The symbol within each box represents the mean for the data range. The mean of all 30 cycles has been normalized to zero to clearly show the variation.

In each cycle, the data range is consistently below 0.10 μm which translates to 3σ values below the repeatability specification of 0.10μm. These results indicate reliable image capture of the optical system and software in both X and Y axes. Results that meet all specifications were obtained on two additional tools.

3.2 Reproducibility

Reproducibility test results for a single measurement session of 30 cycles are also shown in the same box and whisker plots in Figures 4a and 4b. Nearly all the measurement data is within a range of ±0.10μm. This translates to 3σ values well below the specification of 0.30μm. The mean values of each cycle, with the exception of cycle one, are all within ±0.10μm. These results, combined with similar results obtained from two additional UltraMet 100 DSA measurement tools, indicate a consistent ability to exceed the system specifications.

3.3 Reproducibility Over Time

Reproducibility over time was tested on a second tool in five measurement sessions over a 12 day period. The results are shown in Figure 5. The data was collected using the previously described method. Each box and whisker set includes all 30 reproducibility cycles taken each of the five days. The mean of all five measurement sessions has been normalized to zero to more clearly show the variations. Similar data from a second tool shows that all of the measurement data falls within ±0.15μm. The range of mean values of each measurement session is only 0.10μm. Y-axis results are not shown since they are similar to the X-axis.

3.4 Accuracy

A comparison of the pair of chrome target features on the artifact shows only a 2 nm measurement difference in the X axis, and an 80 nm difference in Y. The measurement repeatability of chrome targets on a clear background on the LMS 2020 is within a range of 15 nm to 20 nm based on 20 measurement scans repeated for 20 cycles over 4 days. Measurement repeatability of the UltraMet was demonstrated at less than 100 nm. The measurement difference between the two metrology systems is within the signal to noise level of the design of the UltraMet 100.

4.0 CONCLUSIONS

The UltraMet 100's overall performance in terms of accuracy and its precision, is sufficient to meet current and future dual side alignment metrology requirements of the nanotechnology and MEMS markets. The precision to tolerance ratio (P/T) for this tool meets alignment tolerances in the processes typically employed in current applications and those expected in the near future. Second and third generations of the UltraMet will include substrate handling automation, 200mm wafer handling capability, and improved measurement capability for tighter DSA alignment specifications expected to be required in the future.

Future work will involve tool to tool performance characterization and working on establishing industry wide reference standards for determining DSA metrology tool accuracy.

5.0 ACKNOWLEDGEMENTS

The authors would like to thank Kevin Riddell for his contribution in LMS 2020 measurements and data collection. We would also like to thank Yehuda Pashut (Owens Design) for his assistance with measurement and data analysis, and HTA Photomask for accuracy artifact processing.

6.0 REFERENCES

1. R. Reif, A. Fan, K. Chen, S. Das, "Fabrication Technologies for Three-Dimensional Integrated Circuits", *International Symposium on Quality Design Proceedings*, IEEE, 2002.

2. L. Chen, H. Neves, T. Kudrle, "Process Development for the Integration of MEMS and RF Devices", *Cornell Nanofabrication Users Network*.

3. C. Cheng, J. Hu, Y. Lai, H. Wang, C. Cheng, "Monolithic thermal ink jet print head combining anisotropic etching and electroplating", *Input/Output and Imaging Technologies II Proceedings*, SPIE **4080**, 2000.

4. M. van den Brink, C. de Mol and R. George, "Matching Performance for Multiple Wafer Steppers Using an Advanced Metrology Procedure," *Integrated Circuit Metrology, Inspection and Process Control II Proceedings,* SPIE **921** (1988).

5. *Metrology Tool Gauge Study Procedure for the International 300 mm Initiative (I300I)*, Technology Transfer # 97063295A-XFR.

6. *Evaluating Automated Wafer Measurement Instruments*, Sematech, Technology Transfer 94112638A-XFR.

7. K. Hoshi, E. Kawamura, H. Morohoshi, H. Ina, T. Fujimura, H. Kurita, J. Seligson, "TIS-WIS interaction characterization on overlay measurement tool", *Metrology, Inspection, and Process Control for Microlithography XVI Proceedings*, SPIE **4689**, 2002.

8. http://www.cognex.com/products/PatMax/default.asp, January 2004

9. W. Flack, G. Flores, T. Tran, "Application of Pattern Recognition in Mix-and-Match Lithography", *Optical/Laser Microlithography VIII Proceedings*, SPIE **2440**, 1995.

10. Contributed by B. Smithgall, Image Labs International, January 2004

11. G. Box, W. Hunter, J. Hunter, "Statistics for Experimenters," *John Wiley & Sons* (1978).

Parameter	NanoTech 160
Reduction factor	1X
Wavelength (nm)	390 - 450
Numerical aperture (NA)	0.31
Partial coherence (σ)	0.58
Wafer plane irradiance (mW/cm^2)	1200

Table 1: Optical specifications of the NanoTech 160 stepper used in this study.

Process Step	Parameters	Equipment
HMDS	Vapor Prime, 20 minutes at 150°C	YES oven
Frontside Coat: Shipley S1808	2750 RPM for 30 seconds	ACS200 track
Softbake	Hotplate, 30 seconds at 105°C	ACS200 track
Backside Coat: Shipley S1808	2750 RPM for 30 seconds	ACS200 track
Softbake	Convection Bake, 30 minutes at 105°C	Blue-M Oven
Exposure Frontside (first layer)	Focus: 0 Exposure: 100 mJ/cm^2	NanoTech 160
Develop: Arch OPD262	1 minute puddle, DI water rinse	ACS200 track
Exposure Backside (second layer)	Focus: 0 Exposure: 100 mJ/cm^2	NanoTech 160
Develop: Arch OPD262	1 minute puddle, DI water rinse	ACS200 track
Hardbake	Convection Bake, 30 minutes at 110°C	Blue-M Oven

Table 2: Process conditions for Shipley S1808 for 1.2 µm thickness on silicon substrates.

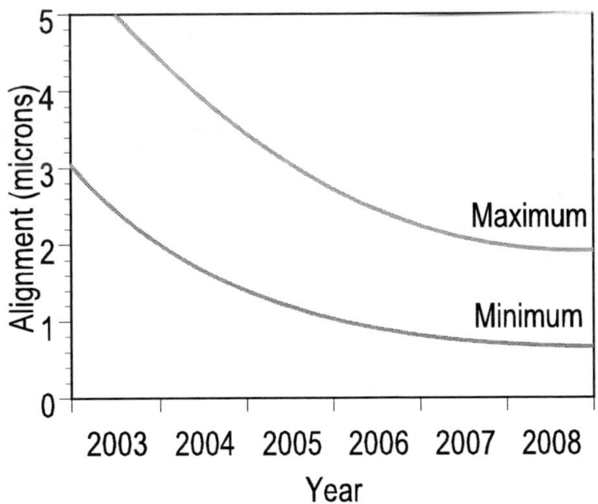

Figure 1: DSA overlay requirements as a function of time. The maximum and minimum curves represent the range of customer DSA requirements.

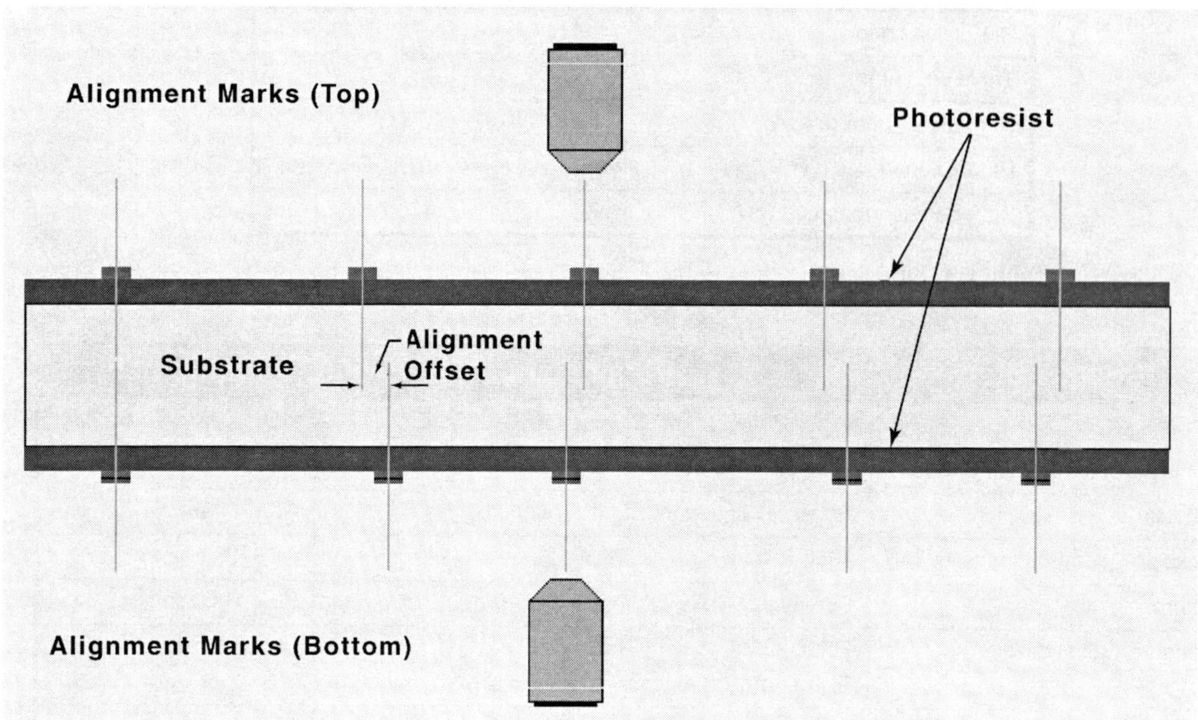

Figure 2: Cross section of a substrate with front and back side patterns in photoresist. The DSA metrology is performed using top and bottom microscopes. The alignment offset is determined by comparing alignment targets on the top and bottom of the wafer.

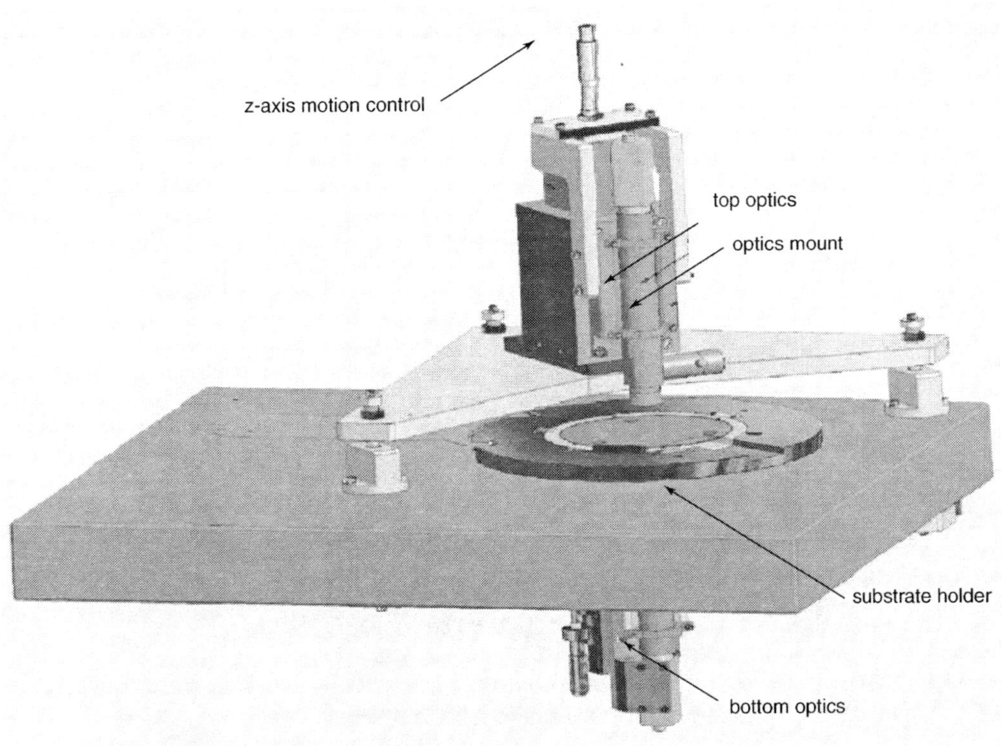

Figure 3: Drawing of UltraMet 100 optical and stage subsystems.

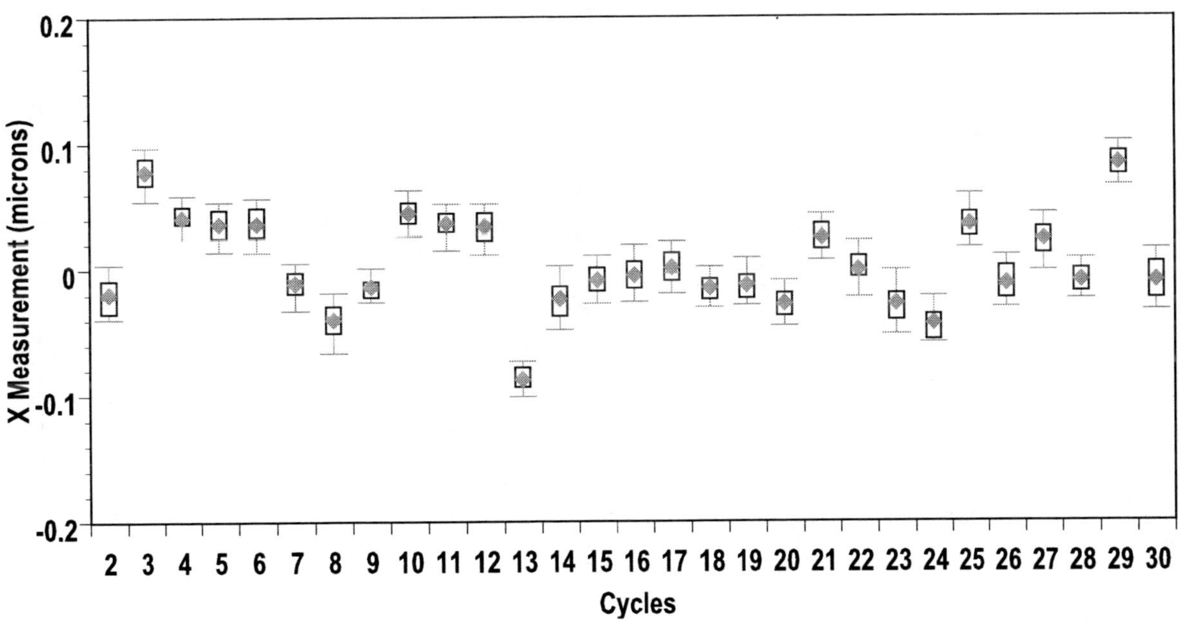

Figure 4a: Box and whisker plot of X reproducibility over thirty wafer cycles. Each cycle consists of thirty repeatability measurements as shown by the box (25 and 75% of data) and whiskers (10 and 90% of data).

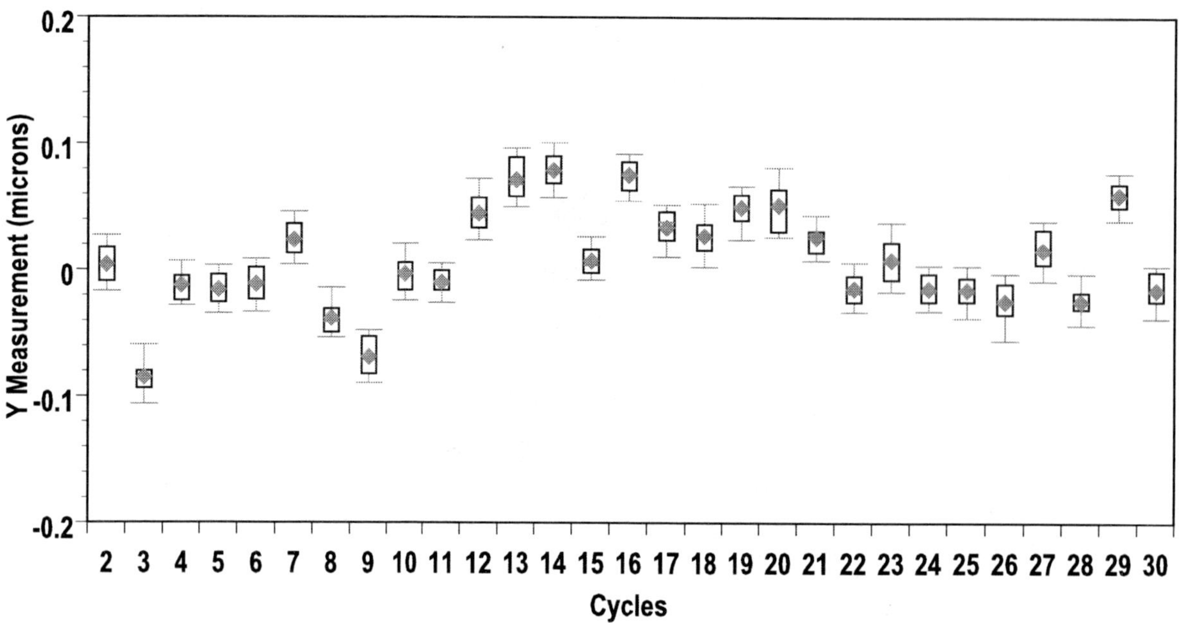

Figure 4b: Box and whisker plot of Y reproducibility over thirty wafer cycles. Each cycle consists of thirty repeatability measurements as shown by the box (25 and 75% of data) and whiskers (10 and 90% of data).

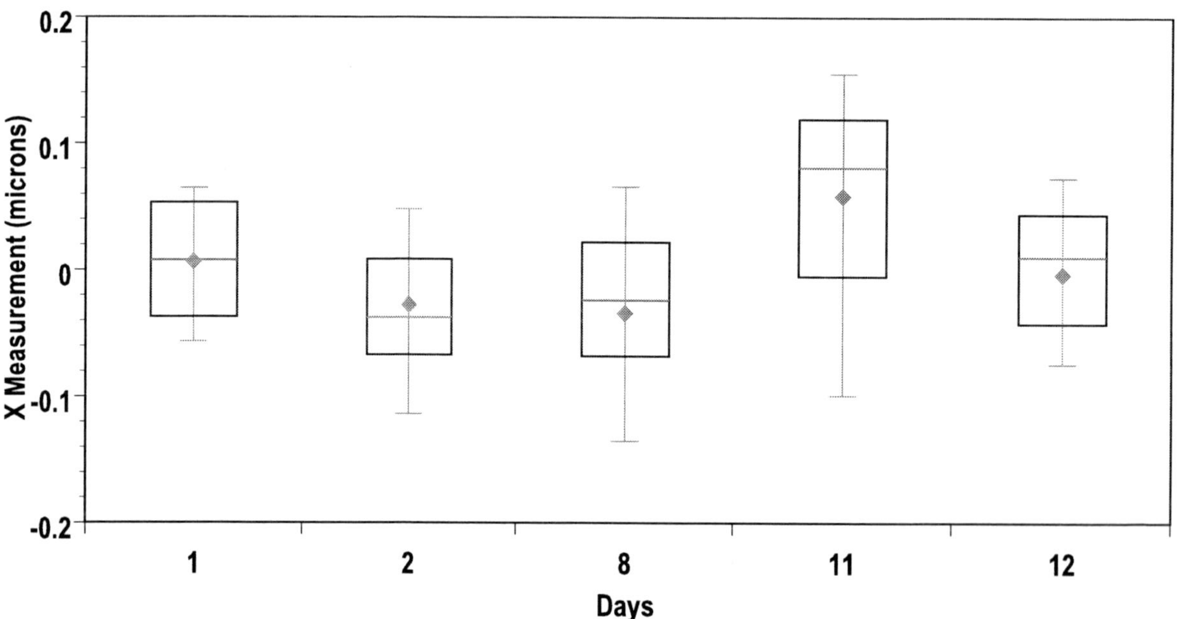

Figure 5: Box and whisker plot of X reproducibility over five days. Each cycle consists of thirty reproducibility cycles as shown by the box (25 and 75% of data) and whiskers (10 and 90% of data).

Novel Microfabrication Techniques for Highly Specific Programmed Assembly of Nanostructures

Balaji Kannan[1] and Arun Majumdar[1,2]

Chemically synthesized nanostructures such as nanowires[1], carbon nanotubes[2] and quantum dots[3] possess extraordinary physical, electronic and optical properties that are not found in bulk matter. These characteristics make them attractive candidates for building subsequent generations of novel and superior devices that will find application in areas such as electronics, photonics, energy and biotechnology. In order to realize the full potential of these nanoscale materials, manufacturing techniques that combine the advantages of top-down lithography with bottom-up programmed assembly need to be developed, so that nanostructures can be organized into higher-level devices and systems in a rational manner. However, it is essential that nanostructure assembly occur only at specified locations of the substrate and nowhere else, since otherwise undesirable structures and devices will result. Towards this end, we have developed a hybrid micro/nanoscale-manufacturing paradigm that can be used to program the assembly of nanostructured building blocks at specific, pre-defined locations of a chip in a highly parallel fashion. As a prototype system we have used synthetic DNA molecules and gold nanoparticles modified with complementary DNA strands as the building blocks to demonstrate the highly selective and specific assembly of these nanomaterials on lithographically patterned substrates.

Two kinds of lithographic patterning processes are developed for nanomaterial assembly. The general microfabrication sequence is outlined in figure 1. In the first method a 25 nm thick gold film is evaporated onto a single crystal silicon wafer. The film is then patterned using standard photolithography and wet etching to define locations on the substrate where nanostructures will be assembled. The chips are cleaned and the silicon surface made hydrophilic by immersing them in a piranha solution (7:3 of concentrated sulfuric acid and 30% solution of hydrogen peroxide). In the second kind of process, the silicon wafer is initially covered with a monolayer of the chemical Aminopropyltrimethoxysilane (APS) using a vapor-phase deposition technique. This monolayer presents amine (NH_2) chemical groups on the surface of the substrate, which can be used to attach various types of molecules to the silicon wafer. Patterns of APS are formed by optical lithography, etching the monolayer using oxygen plasma, dissolving the native oxide in HF (hydrofluoric acid) and allowing it to form again. At this stage the chips have patterns of APS with photoresist on top of them, surrounded by native silicon oxide. Following this, the chips are treated with oxygen plasma to make the silicon hydrophilic and then baked to remove traces of water. From this point on, chips with gold and APS patterns are processed in the same fashion to assemble nanostructures.

The regions of the silicon substrate surrounding the patterns are then passivated or made inert with a layer of poly (ethylene glycol) (PEG) silane. The function of the PEG layer is to prevent the non-specific adsorption of various molecules and nanostructures on the non-patterned parts of the chip, viz. the silicon regions surrounding the patterns. Though it is well known[4] among biochemists that a PEG surface coating exhibits ultra-low non-specific binding towards a wide range of proteins, its use in nanomaterial systems has not been explored so far. In the case of APS patterns, the chips are additionally washed in acetone to remove the photoresist covering the APS monolayer. The assembly process begins with the attachment of single-stranded DNA molecules to the gold patterns and the APS patterns using standard covalent chemistry techniques. Nanoparticles modified with complementary DNA then assemble on this first layer of DNA patterns through Watson-Crick base-pairing. Since each nanoparticle has multiple DNA strands attached to its surface, this process can be repeated many times over to build a multilayer structure of gold nanoparticles. The results of the self-assembly process are shown in figure 2, for both gold-patterned and APS-patterned chips. For comparison we have also presented the results when no surface passivation is used on the surrounding silicon during nanoparticle assembly.

It is seen from figure 2 that throughout the process, assembly occurs exclusively on the metal or chemical patterns and nowhere else because the surrounding silicon is modified with highly inert PEG groups. In fact the inertness of the PEG groups is so extraordinary that there is virtually *zero* non-specific adsorption of nanostructures onto the region surrounding the patterns. In addition it is seen that the second layer of gold nanoparticles attach exclusively to the first nanoparticle layer, since the first and second layers have complementary DNA molecules. Our programmed assembly process thus demonstrates very high specificity and selectivity due to the use of complementary DNA and, more significantly, the PEG surface coating. We are not aware of any coating technique other than PEG that demonstrates such exceptional specificity as seen in our research. In addition, since the rejection capabilities of the PEG layer are purely physical (entropic) in origin[5] it can be employed in conjunction with a variety of other nanomaterials. Thus, this relatively simple but versatile fabrication technique paves the way for seamless integration of chemically synthesized nanostructures with conventional IC fabricated devices and systems.

1. Duan, X. F., Lieber, C. M., *Advanced Materials*, 2000, 12 (4), 298
2. Cassell, A. M., Raymakers, J. A., Kong, J., Dai, H. J., *J Physical Chemistry* B, 1999, 103 (31), 6484
3. Alivisatos, A. P., *Science*, 1996, 271 (5251), 933
4. Zdyrko, B., Klep, V., Luzinov, I., *Langmuir*, 2003, 19, 10179
5. Hermans, J., *Journal of Chemical Physics,* 1982, 77 (4), 2193

[1] Department of Mechanical Engineering, University of California, Berkeley, CA 94720
[2] Materials Sciences Division, Lawrence Berkeley National Laboratory, Berkeley, CA 94720

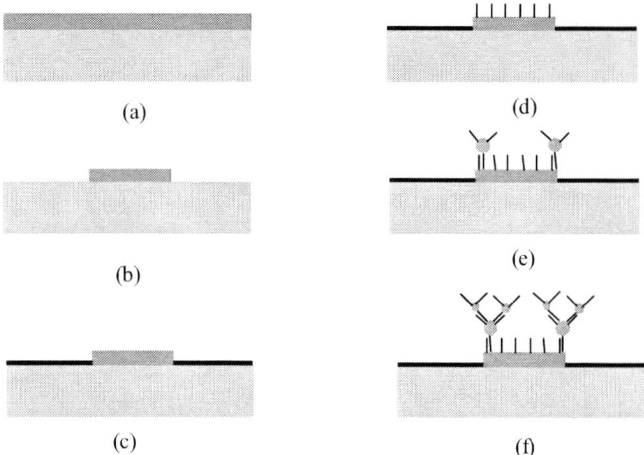

Figure 1. (a) Deposition of Au film on silicon (Si) substrate; (b) Patterning of the Au/APS film by photolithography and wet-etching; (c) Functionalization of the surrounding Si substrate by PEG-silane; (d) Selective assembly of single stranded thiolated DNA on Au /APS patterns; (e) Assembly of first layer of DNA-conjugated Au nanoparticles on patterns; (f) Assembly of second layer of nanoparticles on first nanoparticle layer

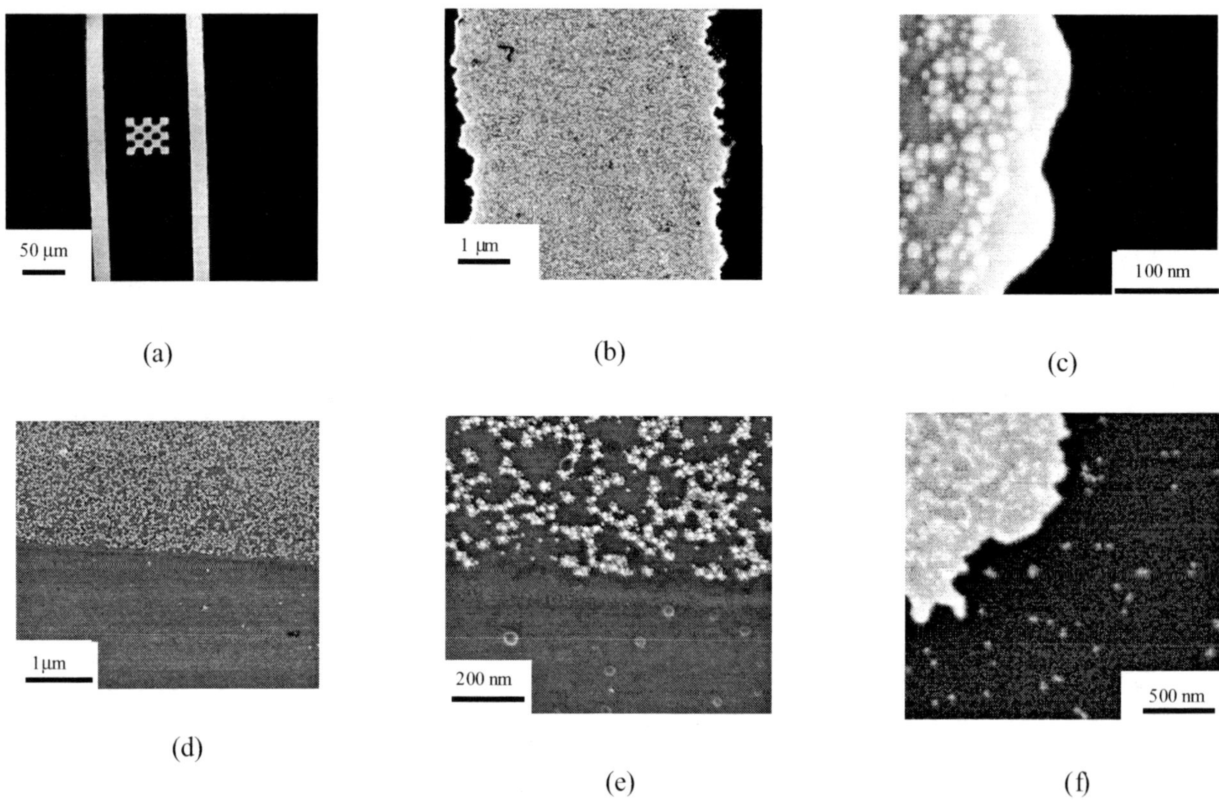

Figure 2. (a) Au or APS patterns on silicon substrate (b) and (c) 2 layers of gold nanoparticles (15 nm followed by 10 nm) on gold patterns surrounded by PEG silane modified silicon (d) and (e) 2 layers of gold nanoparticles on APS patterns surrounded by PEG silane modified silicon (f) 1 layer of gold nanoparticles on gold patterns surrounded by unmodified silicon

114

NANO2004-46056

LASER NANO-MACHINING USING NEAR-FIELD OPTICS

Haseung Chung
Mechanical Engineering Department
University of Michigan
Ann Arbor, MI 48109-2125

Katsuo Kurabayashi
Mechanical Engineering Department
University of Michigan
Ann Arbor, MI 48109-2125

Suman Das
Mechanical Engineering Department
University of Michigan
Ann Arbor, MI 48109-2125

INTRODUCTION

A near-field optical technique, using a new type of solid immersion lens (SIL), has been developed and applied to various areas, for example, high-density optical storage, near-field-scanning-optical-microscope probes, photolithography. Solid immersion microscopy offers a method for achieving resolution below the diffraction limit in air with significantly higher optical throughput by focusing light through a high refractive-index SIL held close to a sample [1]. The minimum resolution of a focusing system is inversely proportional to numerical aperture (NA), where NA=n sinθ, θ is the maximum angle of incidence, and n is the index of refraction at the focal point. Light with vacuum wavelength λ can be focused by an aberration-free lens to a spot whose full width at half maximum (FWHM) is $\lambda/(2\ NA)$ in the scalar diffraction limit, equivalent to Sparrow's criterion for spatial resolution. In a medium of refractive index n, the effective wavelength is $\lambda_{eff}=\lambda/n$ and corresponding effective numerical aperture is NA_{eff} =$n^2\sin\theta$. When a SIL is used, improvements in NA_{eff} and spatial resolution are proportional to the refractive index of the SIL material. Fletcher et al. demonstrated imaging in the infrared with a microfabricated SIL [1, 2]. Baba et al. analyzed the aberrations and allowances for an aspheric error, a thickness error, and an air gap when using a hemispherical SIL for photoluminescence microscopy with submicron resolution beyond the diffraction limit [3]. Terris et al. developed and applied a SIL-based near-field optical technique for the writing and reading domains in a magneto-optic material [4]. Song et al. proposed the new concept of a SIL for high density optical recording using the near-field recording technology [5]. In this paper, we propose a sub-micron scale laser processing technique with spatial resolution beyond the diffraction limit in air using near-field optics. Our goal is to eventually develop a massively parallel nano-optical direct-write nano-manufacturing technique.

EXPERIMENTAL

Figure 1 is a schematic of the sample setup for conducting near-field nanomachining experiments. Three different SIL sizes, 2mm, 4mm and 6mm were tested at different SIL-substrate distances in the 20-70nm range using oxide pedestals that were grown and patterned onto 500μm thick silicon wafers.

In our study, a λ=532nm Q-switched Nd:YAG laser is used and is controlled by an external trigger input. The diameter of the TEM$_{oo}$ beam from the cavity is magnified 5 times by a beam expander and then focused on a silicon wafer through a micro focusing objective that has a 4μm theoretical spot size in air. The converging laser beam is transmitted through a BK7 SIL placed on 50nm SiO$_2$ nano-spacer patterns created on top of the 500 μm thick silicon wafer as shown in figure 1.

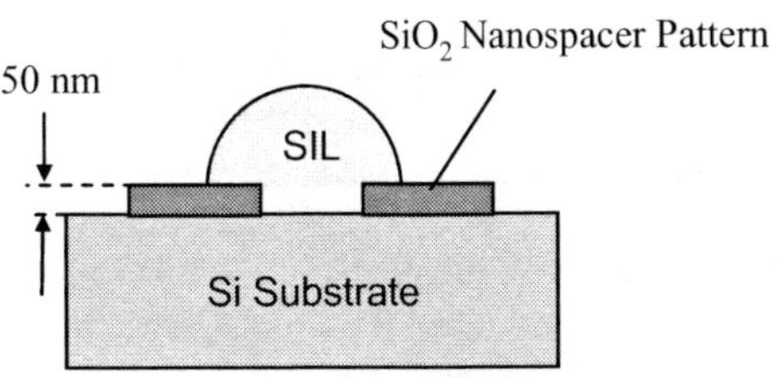

Figure 1. Schematic description of sample

When only a microfocus objective is used without a SIL, the numerical aperture of the system is the numerical aperture of the objective and is 0.4. However, the NA of the system is increased to NA_{eff}=0.9237 when a BK7 SIL with n=1.5196 is interposed between the objective and the silicon wafer. Therefore, placing a BK7 SIL yields near diffraction limited performance with respect to focusing and a better spatial resolution with respect to processed features is expected.

RESULTS AND DISCUSSION

Figures 2a and 2b are scanning electron microscopy (SEM) images of laser processing experiments conducted at 1.04 Watts peak pulse power, 10 kHz repetition rate, and 50μsec pulse width respectively without a SIL and with a 2mm diameter BK7 SIL.

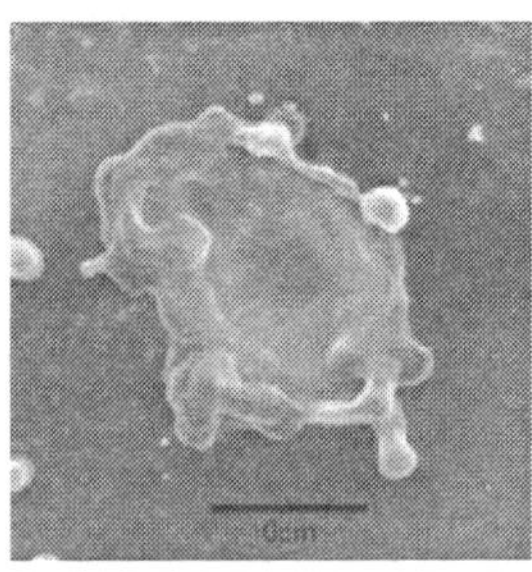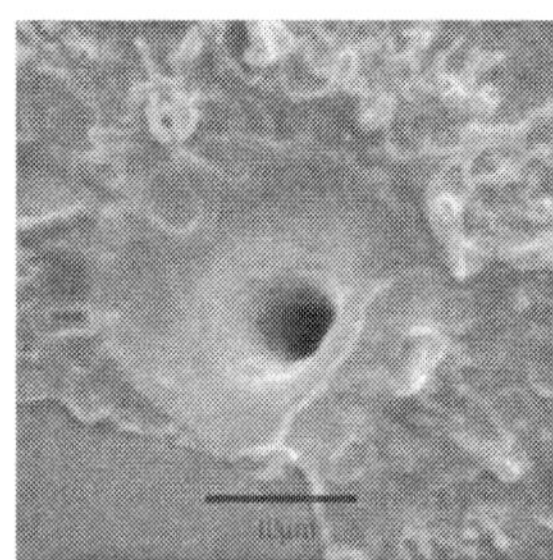

Figure 2. Effect of SIL on processed feature size (a) without SIL (b) with 2mm BK7 SIL

As can be seen, figure 2b shows a sharper laser processed feature due to a more focused laser spot than the one in figure 2a. There is a distinct central hole and overall feature size is bigger in figure 2a. This difference arises from the increased effective numerical aperture NA_{eff}=0.9237 obtained by interposing the BK7 SIL between the mirofocus objective and the silicon substrate, resulting in improved near-field focusing of the laser and consequently finer spatial resolution.

Several experiments were conducted at various processing conditions to investigate the effects of varying incident laser power (with average pulse power less than 1W), pulse repetition rate, pulse width, number of pulses and SIL diameter on processed feature size and resolution. Figure 3 shows the average diameter of the hole as a function of peak power obtained by conducting three experiments at each peak power setting. We observe that the diameter of the central hole increases as peak pulse power is increased keeping pulse repetition rate, pulse width, number of cycle, and SIL size fixed. Similarly, the average diameter of the hole as a function of pulse repetition rate, pulse width, number of pulses and SIL diameter was investigated. We conclude from our experimental results that the feature size obtainable by SIL-based near-field laser machining of silicon using a 532nm Q-switched laser is proportional to the peak pulse power, number of pulses, pulse width, and SIL diameter keeping other conditions fixed. The laser machined feature is also inversely proportional to the pulse repetition rate. Therefore, smaller laser machined features are attainable using a Q-switched laser with low peak power, fewer pulses (preferably single pulse), shorter pulse width, smaller SIL diameter and large pulse repetition rate.

Furthermore, SILs made of materials with higher refractive index can provide a larger NA_{eff} resulting in smaller focused near-field spot size and also smaller processed feature size. For this purpose, SILs made of Sapphire (n=1.7718, NA_{eff}=1.2557) and LaSFN9 (n=1.8590, NA_{eff}=1.3824) with diameters smaller than 2mm are being investigated.

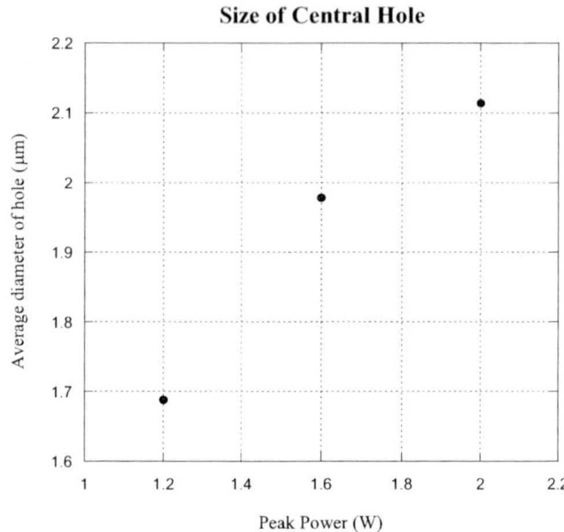

Figure 3. Average diameter of central hole vs. peak power

ACKNOWLEDGMENTS
This work was supported by National Science Foundation Grant 0210240.

REFERENCES
1. Fletcher, D.A., Crozier, K.B., Quate, C.F., Kino, G.S., and Goodson, K.E., 2000 "Near-field infrared imaging with a microfabricated solid immersion lens", Applied Physics Letters, 77 (14), pp. 2109-2111.

2. Fletcher, D.A., Goodson, K.E., and Kino, G.S., 2001 "Focusing in microlenses close to a wavelength in diameter", Optics Letters, 26 (7), pp. 399-401

3. Baba, M., Sasaki, T., Yoshita, M., and Akiyama, H., 1999 "Aberrations and allowances for errors in a hemisphere solid immersion lens for submicron-resolution photoluminescence microscopy", Journal of Applied Physics, 85 (9), pp. 6923-6925.

4. Terris, B.D., Mamin, H.J., Rugar, D., Studenmund, W.R., and Kino, G.S., 1994 "Near-field optical data storage using a solid immersion lens", Applied Physics Letters, 65 (4), pp. 388-390.

5. Song, T.S., Kwon, H.D., Yi, M.D., Park, N.C., and Park, Y.P., 2002 "Optical recording system using an integrated solid immersion lens", Microsystem Technologies, 8, pp. 169-173.

116

NANO2004-46060

Micro contact printing of DNA molecules

H. Zhou, K. Ma, G. Jia, J. Zoval and M. Madou
Henry Samueli School of Engineering,
University of California, Irvine

The development of DNA sensors has attracted substantial research efforts. Such devices could be used for the rapid identification of pathogens in humans, animals, and plant; in the detection of specific genes in animal and plant breeding; and in the diagnosis of human genetic disorders. The first step to fabricate the DNA sensors is the probe immobilization on the suitable substrate. Traditionally, the DNA probes are spotted on the substrate while the technique hardly controlled the small pattern and surface density of DNA probes. The main challenge here is to achieve probe layer uniformity and the nature of the probe layer itself in few micron and sub-micron feature range.

Microcontact printing is a reliable technique for directly defining features in thin films. It is a practical and convenient means of introducing alkanethiols functionalities onto gold surface. Alkanethiols adsorb onto the gold surface spontaneously. The thiol groups chemisorb onto the gold surface through the formation of a thiol-gold bond to produce an ordered monolayer. Patterned self-assembled monolayer (SAM) can be subsequent modified as linker of DNA probe. The ordered DNA self-assembled monolayer probe with higher integrity can be achieved which provided the more reproducible and sensitive DNA arrays.

We have developed a novel scheme for fabrication of DNA arrays on gold substrates based on the microcontact printing method. As shown in figure 1, we used 11-mercaptoundecanoic acid as the ink material. An elastomeric stamp, fabricated from poly(dimethylsiloxane) (PDMS), was used to deliver the mercaptoundecanoic acid to the surface of an evaporated gold film. The SAM patterned gold surface was then activated by 1-ethyl-3-(3-dimethylaminopropyl) carbodiimide hydrochloride (EDC) and oligonucleotide with an amine linker was added afterwards to form an amide linkage. It was confirmed that a well-defined layer of DNA molecule was successfully immobilized on the gold substrate using this method. Besides, as shown in figure 2, we tested the DNA hybridization capability for the probes with this kind of immobilization technique. We will also show that by this method, DNA can be easily immobilized on gold with high resolution in the submicrometer range. The method allows for efficient attachment of DNA strands to gold substrates and this new technique could be utilized for rapid preparation of DNA-assays and genetic biochips.

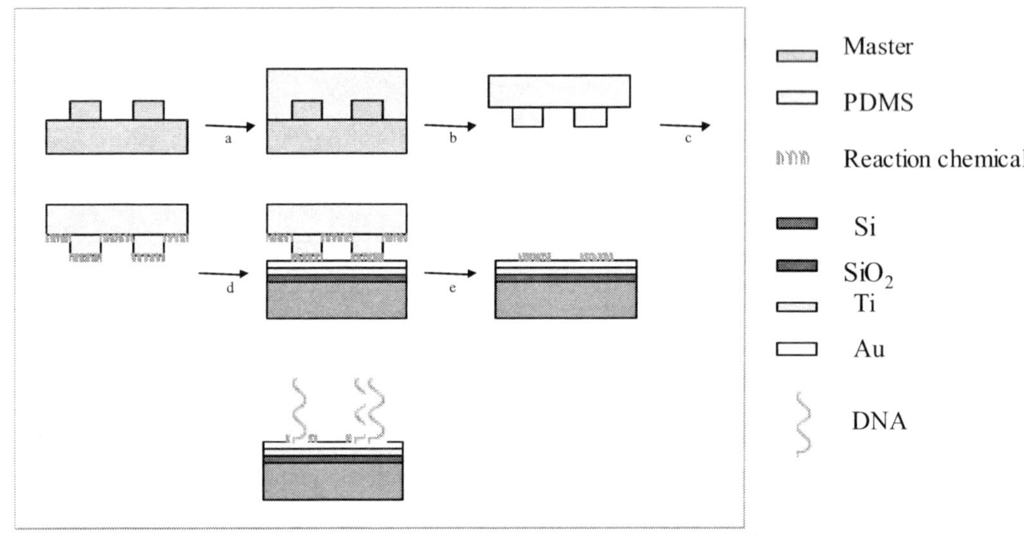

Fig.1   Schematic diagram showing the process of immobilizing DNA on gold substrate by microcontact printing method.

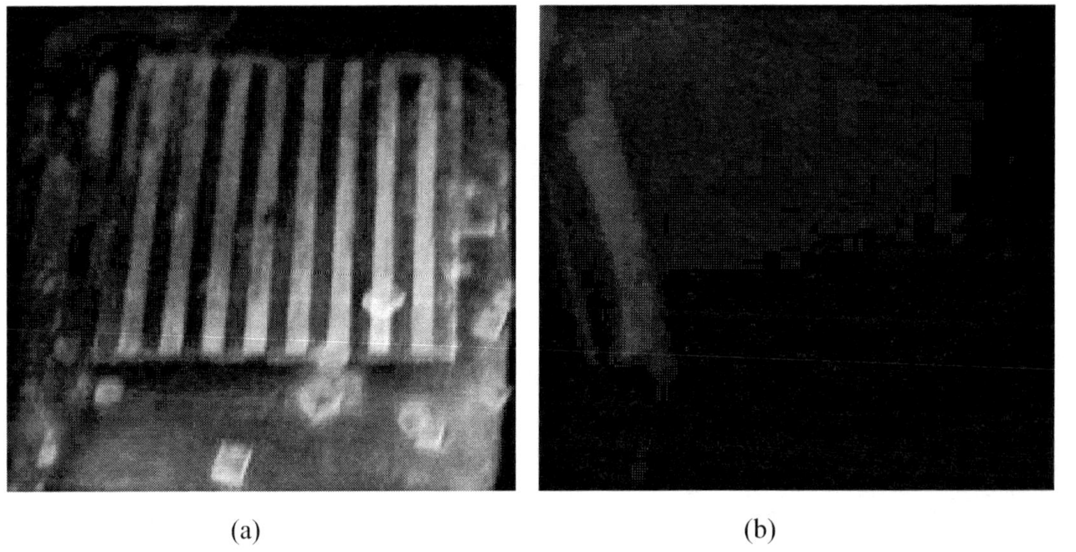

(a)   (b)

Fig. 2 DNA hybridization fluorescent image by the microcontact printing technique. (a) specific hybridization. (b) non-specific control experiment.

NANO2004-46067

AN INTEGRATED MEMS SYSTEM FOR IN-SITU MECHANICAL TESTING OF NANOSTRUCTURES

Shaoning Lu, Jaehyun Chung, Dima Dikin,Junghoon Lee, and Rodney S. Ruoff

Department of Mechanical Engineering, Northwestern University
2145 Sheridan Road, Evanston, IL 60208-3111
*r-ruoff@northwestern.edu

ABSTRACT

We present a hybrid nano-electromechanical system for study of the mechanics of nanostructures. The system has a testing platform based on a deep reactive ion-etched high aspect ratio MEMS device. A new approach has been developed with top-down manufacturing of the micro-device and bottom-up post-fabrication assembly of samples (nanostructures) to be tested. A process that minimizes chemical or physical damage of the sample is used to integrate suspended nanowires/nanotubes into the system. The system provides nanoscale resolution of displacement and force. The device is used in an SEM and is being tested for *in situ* experiments on various nanowires or nanotubes.

EXPERIMENT

Multi degree-of-freedom robotic manipulators have been used for study of the mechanics of CNTs [1,2]. However, this multi DOF manipulator is larger than needed for *in-situ* testing of nanostructures in a scanning electron microscope (SEM) much too large for insertion into a transmission electron microscope (TEM). Microfabricated miniature stressing stages have been suggested for nanostructure testing, but have been limited in the displacement, force, their structural stability, and materials that can be tested [3]. We present here a thermally actuated, high-aspect ratio MEMS testing stage that realizes a high out-of-plane stiffness and fine motion control with step size resolution of 30 nm. Figure 1 shows an SEM image of one such device.

The fabricated stages were characterized inside an SEM (LEO1525) and also with an optical profilometer in ambient. The device has a working range from tens of nanometers to 10 micrometers. Displacements as small as 30 nm for 10 mA dc input current increments were realized. The height difference (offset) between the moving and fixed platforms was less than 40 nm over the entire working range of the device for the input power range studied. A small (ideally, zero) offset is critical for measuring the mechanics of nanostructures.

Studies of mechanics of nanostructures can benefit from suspended specimens, either for tensile or compressive loading, or vibrating the sample. We have developed a post-fabrication assembly process. Figure 2 shows the general concepts; we suspend individual CNTs in a desired location with this approach. A design modification on the MEMS device has been made to allow for the precise deposition of only one CNT on top of the platforms that are used to mechanically load samples suspended between them. Figure 3 shows a typical result on a recently fabricated device with a single T-CNT (*templated* carbon nanotube) with diameter ~300nm that has been deposited across a 6-μm gap which is 80-μm deep and has an aspect ratio of 13:1. In situ SEM testing of the current set of devices, fabricated in June, 2004 at the Cornell Nanofabrication is now underway.

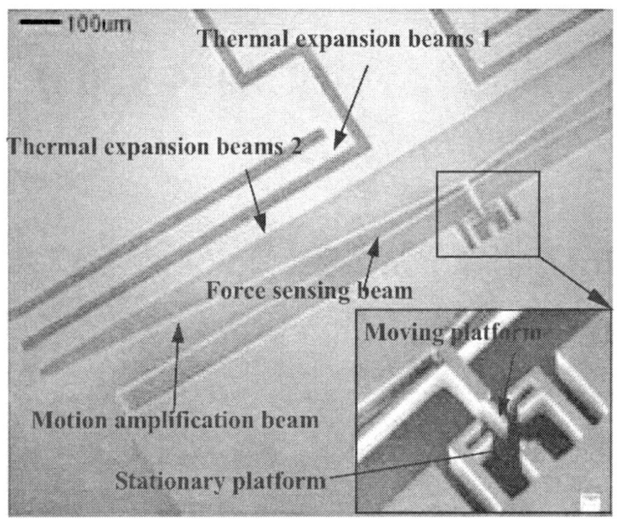

Fig. 1 SEM image of the device. A force-sensing beam is attached to the fixed side and connected to the moving platform at its center. The inset is a magnified view of the gap region. The gap between the moving platform and the stationary platform is 20 µm.

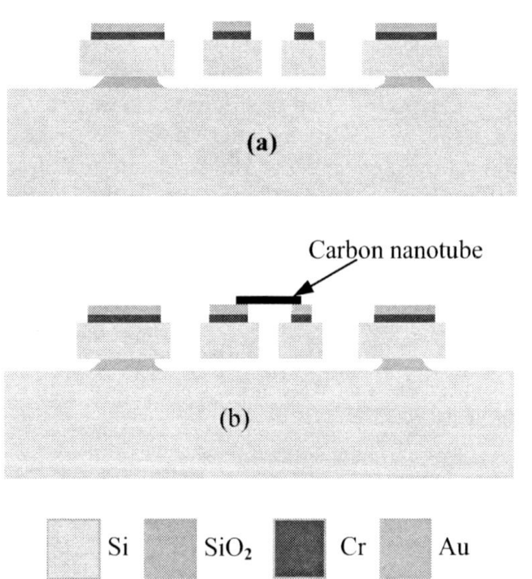

Fig. 2 (a) Bulk micromachining of the device by using DRIE on a SOI wafer; (b) integration of a CNT on top of the device for tensile testing.

Fig 3 A templated carbon nanotube on the MEMS device.

ACKNOWLEDGMENTS

We gratefully acknowledge the Office of Naval Research *Mechanics of Nanostructures* grant (award No. N000140210870), and the NASA University Research, Engineering and Technology *Institute on Bio Inspired Materials (BIMat)* under award No. NCC-1-02037. This work was performed in part at the Cornell Nano-Scale Science & Technology Facility (a member of the National Nanofabrication Users Network) which is supported by the National Science Foundation under Grant ECS-9731293, its users, Cornell University and Industrial Affiliates. SEM was performed using the NUANCE facilities at Northwestern University.

REFERENCES

[1] Yu, M. F., Files, B. S., Arepalli, S., Ruoff, R. S., *Physical Review Letters*, 2000. **84**(24): p. 5552-5555.
[2] Yu, M. F., Lourie, O., Dyer, M. J., Moloni, K., Kelly, T. F., Ruoff, R. S., *Science*, 2000. **287**(5453): p. 637-640.
[3] Saif, M. T. A., Zhang, S., Haque, M. A., Hsia, K. J., *Acta Materialia*, 2002. **50**: p. 2779-2786.

120

NANO2004-46072

NEAR-FIELD CALIBRATION OF ISOLATED CARBON NANOTUBE FIELD EMITTERS FOR NANOMANUFACTURING APPLICATIONS

King-Fu Hii[*], **R. Ryan Vallance**[†], **Padmakar D. Kichambare**[*], **and M. Pinar Mengüç** [‡]

[*] Mechanical Engineering, University of Kentucky
015 Ralph G. Anderson Building, Lexington, KY, 40506, USA

[†] Mechanical and Aerospace Engineering, The George Washington University
738 Phillips Hall, 801 22nd St., N.W., Washington, DC, 20052, USA
Tel: +1-202-994-9830; Fax: +1-202-994-0238. Email: vallance@gwu.edu

[‡] Department of Mechanical Engineering, University of Kentucky
269 RGAN Building, Lexington, KY 40506

ABSTRACT

This paper reports the development of an apparatus, technique, and method for calibrating the field emission phenomena's dependence on both the voltage applied between the anode and cathode and the electrodes gap. A precise knowledge of the electrodes gap is required for calibrating field emitters. The I-V characteristic of isolated carbon nanotube field emitter is a strong function of the electrodes gap distance. A consolidated IV curve is obtained by calculating the current density and the local electric field with the field enhancement factor taken into consideration. The field enhancement factor and emitting area are unique for each electrodes gap distance. We also found that the turn-on voltage decreases as the electrodes gap distance decreases.

INTRODUCTION

The phenomenon of electron field emission from carbon nanotubes (CNTs) was suggested as a means for achieving highly concentrated energy density in nanomanufacturing applications [1,2]. Recent simulations by Wong et al. [3,4] suggest that material removal by vaporization is feasible if the electron beam and a supplementary laser transfer sufficient power to the workpiece. For such applications, the tool or cathode can consist of an isolated carbon nanotube attached to a metallic electrode, and the workpiece or anode might be any conductive substrate. The success of such a nanomanufacturing technique requires quantitative assessment in a repeatable and reproducible manner that is adequate for inter-comparison of emitters.

As suggested by prior field emission reports [5,6,7,8], the dependence of the emitted current density on the gap and applied voltage is usually adequately modeled by the Fowler-Nordheim (F-N) equation (with image charge potential correction) given in Eq. (1), where, ϕ and E are the work function in eV and local electric field in V/m, respectively. The constants B, C, and D are 1.5×10^{-6} AeVV^{-2}, 10.4 eV$^{0.5}$, and 6.44×10^{9} VeV$^{-1.5}$m^{-1}, respectively.

$$ J = B \frac{E^2}{\phi} \exp\left(\frac{C}{\sqrt{\phi}} - \frac{D\phi^{1.5}}{E} \right) \qquad (1) $$

Rewriting Eq. (1) for emitted current density, $J = I/A$, and local electric field, $E = \beta V/d$, yields the emitted current given in Eq. (2). Where, A, V, d, and β are emitting area in m^2, applied voltage in V, anode-cathode gap in m, and field enhancement factor, respectively.

$$ I = A \frac{B}{\phi} \left(\frac{V}{d} \right)^2 \beta^2 \exp\left(\frac{C}{\sqrt{\phi}} - \frac{D\phi^{1.5}d}{V} \right) \qquad (2) $$

In most field emission studies, the applied voltage is varied from values below the turn-on voltage (voltage at which the emitted current is 1 nA) up to a maximum voltage limited by either the high voltage power supply or damage to the CNT. Often the apparatus does not permit precise knowledge of the gap distance, so most are conducted at large gaps where that uncertainty in the gap distance is small compared to the nominal gap. This is not reasonable for near-field measurements where the nominal gap is below 50 µm; in such cases, the gap distance must be known and accounted for in the measurements. Furthermore, slight adjustments in the gap distance can have a dramatic influence on the emitted current suggested by the F-N equation.

APPARATUS AND INSTRUMENTATION

Several pieces of apparatus and instruments such as nanopositioning stage, current amplifier, and high voltage source are needed for calibrating field emitters. The nanopositioning stage as illustrated in the Fig. 1 is a linear motion flexural stage actuated by a PZT actuator can deliver up to 20 μm of displacement. A variable gain setting (10^3 to 10^{11} V/A) Keithley current amplifier (Model 428) is used to measure the emitted current from field emitter. And, a low noise and low current leakage high voltage power supply from Bertan (Model 230-03R) with remote control capability is used for applying the biasing voltage between the electrodes.

The anode is a 125 μm diameter optical fiber coated with a thin layer of copper in the radial direction and gold at the end surface. The cathode is an etched tungsten wire (150 μm diameter) with isolated carbon nanotubes attached to the tip. Two quartz substrates with micro Vee grooves are used to align the electrodes. The initial gap separation between the electrodes is adjusted by pushing the cathode with a micrometer stage. The electrical connection to the electrodes is a stainless steel wire that preloads each electrode to its corresponding quartz Vee groove.

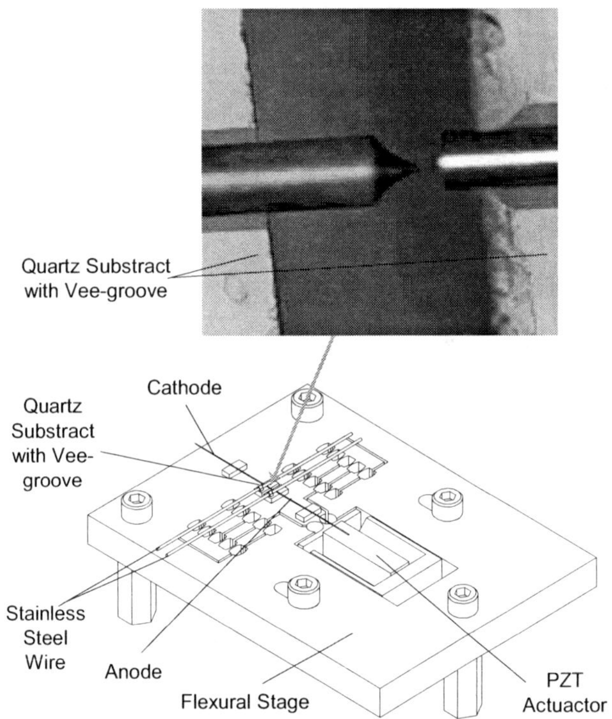

Fig. 1: Apparatus for calibrating field emitter

CALIBRATION PROCEDURE

Two crucial components in calibrating field emitters are the precise knowledge of the initial anode-cathode gap and a few emitted current measurements at their corresponding gap distances. In this work, the initial gap is measured to be 35 μm with a 500x optical microscope. All field emission measurements were performed within a vacuum chamber at base pressure of 1.6×10^{-8} mbar and for a total of 21 gap distances ranging from 15 μm to 35 μm with 1 μm increment. The current amplifier gain setting used is 10^7 V/A. Three current measurements were recorded for the applied voltage varying from values below the turn-on voltage (voltage at which the emitted current is 1 nA) up to the voltage when 1 μA emitted current is detected. The average of three measurements is used to represent the current-voltage (I-V) characteristic of the emitter for that particular gap distance. The emitting area and the field enhancement factor for each I-V curve are calculated from the best-fitted F-N equation to the measured I-V data. The work function of the isolated carbon nanotubes is assumed to be 4.8 eV taken from the typical carbon-based field emitters.

RESULTS AND DISCUSSION

Figure 2 illustrates the current-voltage (I-V) curves for gap distances between 15 μm and 35 μm. The I-V curves shift to the left as the gap distance decreases. The shift in the I-V curves showed that a lower voltage is required to achieve the same amount of emitted current. The turn-on voltage ranges from 1.2 kV to 1.6 kV. For emission currents below 0.1 μA, the emitted current is well modeled by the F-N equation. However, a portion of the I-V curves show a rapid increase in the emission current above 0.1 μA.

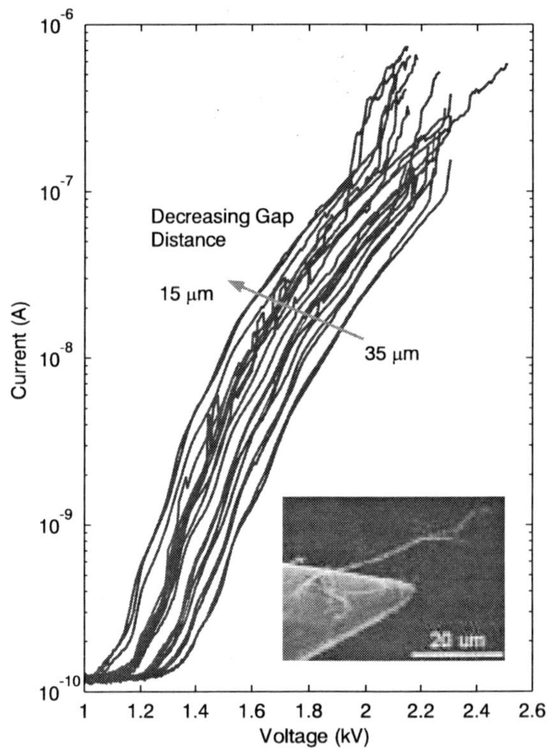

Fig. 2: I-V curves

Figure 3 illustrates the consolidated I-V curve for all the gap distance field emission measurements using the values of the field enhancement factor and emitting area as shown in the Fig. 4 and Fig. 5, respectively, and well modeled by the F-N law (from Eq. (1) - solid line). This consolidated IV curve is obtained by calculating the current density and the local electric field with the field enhancement factor taken into consideration. The turn-on field is ~5 V/nm.

Copyright © 2004 by ASME

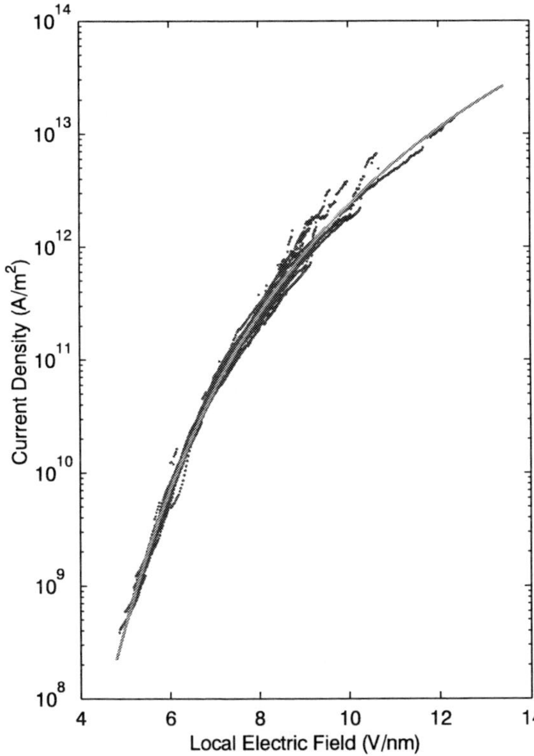

Fig. 3: Consolidated I-V curve and the F-N curve

The field enhancement factor decreases almost linearly from 130 to 75 with decreasing gap distance. However, the emitting area is almost independent of the gap distance. The emitting area is relative large for a spike decrease in the field enhancement factor (e.g. at gap distance equals to 21 µm).

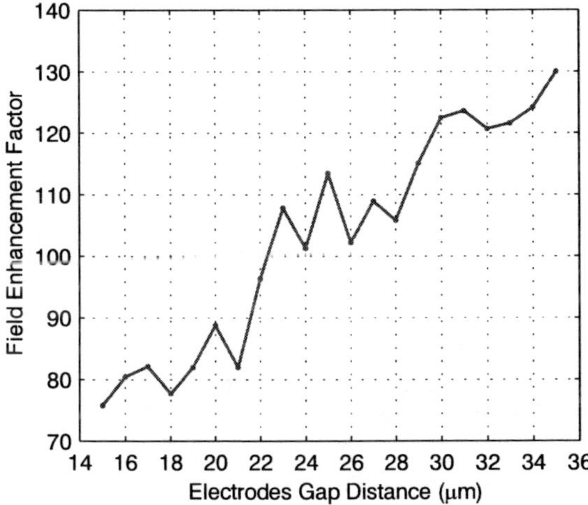

Fig. 4: Calculated field enhancement factor

CONCLUSION

The I-V characteristic of isolated carbon nanotube field emitter is a strong function of the electrodes gap distance. The field enhancement factor and emitting area are unique for each particular gap distance. The I-V characteristic of field emitters maybe samples dependence. For inter-comparison of emitters, field enhancement factor, emitting area, and turn-on field can be used.

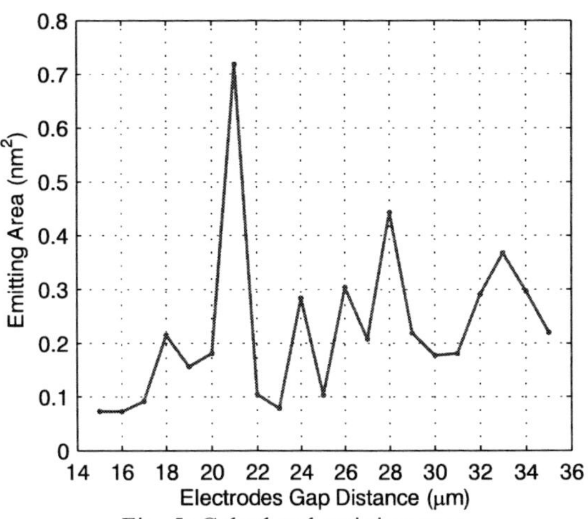

Fig. 5: Calculated emitting area

ACKNOWLEDGMENTS

This work is supported by an NSF Nanoscale Interdisciplinary Research Team (NIRT) award from the Nano Manufacturing program in Design, Manufacturing, and Industrial Innovation (DMI-0210559).

REFERENCES

[1] Trinkle, C.A., Smith, P., Vallance, R.R., Wong, B.T., and Mengüç, M.P., Nov. 17-22, 2002, "Thermal Finite Difference Analysis of Threshold Heating for Nanoscale Machining," Proceedings of the IMECE: International Mechanical Engineering Conference & Exposition, New Orleans, LA.

[2] Trinkle, C.A., Vallance, R.R., Mengüç, M.P., Bah, A., Javed, K., Rao, A.M., and Jin, S., Oct. 2002, "Nanoprobe Concepts for Field Emission Nanomachining," Proceedings of the 2002 Annual Meeting of the American Society for Precision Engineers, St. Louis, MO.

[3] Wong, B.T., Mengüç, M.P., and Vallance, R.R., "Nano-scale Machining Via Electron Beam and Laser Processing," ASME Journal of Heat Transfer.

[4] Wong, B.T., Mengüç, M.P., and Vallance, R.R., "Sequential nano-patterning Using Electron and Laser Beams: A Numerical Methodology," Fourth International Symposium on Radiation Transfer, June 20-25, 2004.

[5] Bonard, J.M., Dean, K.A., Coll, B.F., and Klinke, C., 2002, "Field Emission of Individual Carbon Nanotubes in Scanning Electron Microscope," Physical Review Letters, **89**, pp. 197602.

[6] Bonard, J.M., Salvetat, J.P., Stockli, T., Forro, L., and Chatelain, A., 1999, "Field Emission from Carbon Nanotubes: Perspertives for Applications and Clues to the Emission Mechanism," Applied Physics A: Materials Science & Processing, **69**, pp. 245-254.

[7] Fransen, M., 1999, *Towards High-brightness, Monochromatic Electron Sources*, PhD. Thesis, Delft University of Technology, Delft, The Netherlands.

[8] de Jonge, N., van Druten, N.J., 2003, "Field Emission from Individual Multiwalled Carbon Nanotubes Prepared in an Electron Microscope," Ultramicroscopy, **95**, pp. 85-91.

NANO2004-46074

TRANSPORT IN THERMAL DIP PEN NANOLITHOGRAPHY

Brent A. Nelson, Tanya L. Wright, William P. King
Woodruff School of Mechanical Engineering
Georgia Institute of Technology
Atlanta, GA 30332
Paul E. Sheehan, Lloyd J. Whitman
Naval Research Laboratory
4555 Overlook Ave SW
Washington, D.C. 20375-5320

The manufacture of nanoscale devices is at present constrained by the resolution limits of optical lithography and the high cost of electron beam lithography. Furthermore, traditional silicon fabrication techniques are quite limited in materials compatibility and are not well-suited for the manufacture of organic and biological devices. One nanomanufacturing technique that could overcome these drawbacks is dip pen nanolithography (DPN), in which a chemical-coated atomic force microscope (AFM) tip deposits molecular 'inks' onto a substrate [1]. DPN has shown resolution as good as 5 nm [2] and has been performed with a large number of molecules, but has limitations. For molecules to ink the surface they must be mobile at room temperature, limiting the inks that can be used, and since the inks must be mobile in ambient conditions, there is no way to stop the deposition while the tip is in contact with the substrate. In-situ imaging of deposited molecules therefore causes contamination of the deposited features.

This paper describes *thermal* dip pen nanolithography (tDPN) which uses AFM cantilevers with integrated heaters that were originally developed for thermomechanical data storage by Stanford [3] and IBM [4], but have recently fabricated at Georgia Tech as well, shown in Fig. 1. In tDPN, a molecular ink with a melting temperature above ambient is selected and then coated onto the AFM tips. By heating the cantilever above the ink melting temperature, material can be deposited from the tip onto the surface, as shown schematically in Fig. 2. Our first results use octadecylphosphonic acid (OPA) as our ink, which melts at 99 °C [5]. Figure 3 shows 500nm square features that illustrate temperature modulation of deposition rate. At ambient temperature and at 57 °C, no deposition is observed, showing that the tip does not contaminate the surface at low temperatures.

As the tip temperature reached 98 °C, some deposition occurred, while at a tip temperature of 122 °C, the OPA became more mobile and a more robust deposition occurred. Preliminary results have also shown potential for tDPN of low melting temperature metals such as indium, which could be used for nanoscale soldering. Our hypothesis is that the heat and mass transfer phenomena associated with the deposition are a coupled phenomena, and significant work and modeling remains to be done to learn more about the temperature fields in the tip and substrate.

tDPN is a nanomanufacturing technique that could overcome the limitations to standard DPN because of its ability to locally control the ink deposition process. The ability to locally control deposition creates the possibility of in-situ surface imaging and modification. Additionally, tDPN offers the potential for small-volume manufacturing and prototyping without the use of optical or electron beam lithography. Large arrays of heater-cantilevers have been fabricated [6] with which high throughput manufacturing would be possible. tDPN could also pattern *functional* organic molecules, allowing for direct-write manufacturing.

REFERENCES

[1] Piner, R. D., Zhu, J., Xu, F., Hong, S. and Mirkin, C. A., 1999, "Dip pen nanolithography", Science, **283**, pp. 661-663
[2] Mirkin, C. A., Hong, S. and Demers, L. M., 2001, "Dip-Pen Nanolithography: Controlling Surface Architechture on the Sub-100 Nanometer Length Scale", ChemPhysChem, **2**, pp. 37-39
[3] Chui, B. W., Stowe, T. D., Ju, Y. S., Goodson, K. E., Kenny, T. W., Mamin, H. J., Terris, B. D. and Ried, R. P., 1998, "Low-stiffness silicon cantilever with integrated heaters and

piezoresistive sensors for high-density data storage", Journal of Microelectromechanical Systems, **7**, pp. 69-78

[4] Mamin, H. J. and Rugar, D., 1992, "Thermomechanical writing with an atomic force microscope tip", Applied Physics Letters, **61**, pp. 1003-1005

[5] Kosolapoff, G. M., 1945, "Isomerization of Alkylphosphites .3. The Synthesis of Normal-Alkylphosphonic Acids", Journal of the American Chemical Society, **67**, pp. 1180-1182

[6] Vettiger, P., Cross, G., Despont, M., Drechsler, U., Durig, U., Gotsman, B., Haberle, W., Lantz, M., Rothuizen, H., Stutz, R. and Binnig, G., 2002, "The "millipede" - nanotechnology entering data storage", IEEE Transactions on Nanotechnology, **1**, pp. 39-55

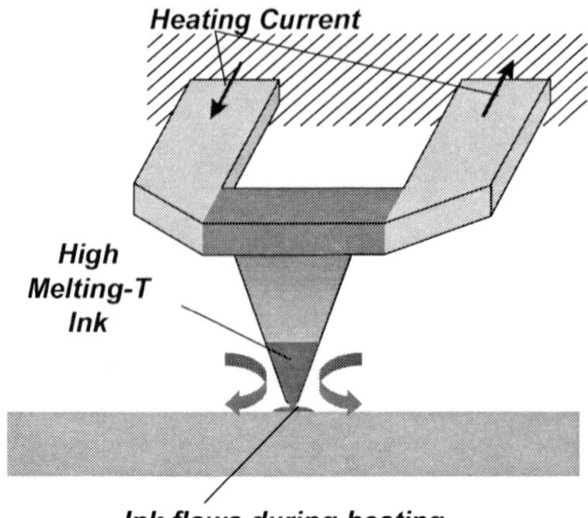

Figure 2: Illustration of Thermal Dip Pen Nanolithography (tDPN). At high temperatures, the ink melts and is transferred from the cantilever tip onto a substrate.

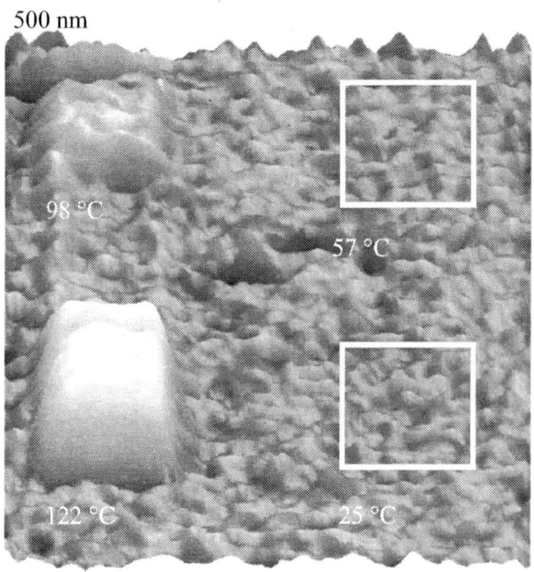

Figure 3: 3D image of mica scanned with a heated AFM tip for 256 seconds in each of four 500 nm squares. The tip temperature is shown for each of the four scans. No deposition is observed from the two low temperature scans. The scan at 98° C resulted in light deposition, while robust deposition occurred at a tip temperature of 122°C. The scan size is 2.5 μm.

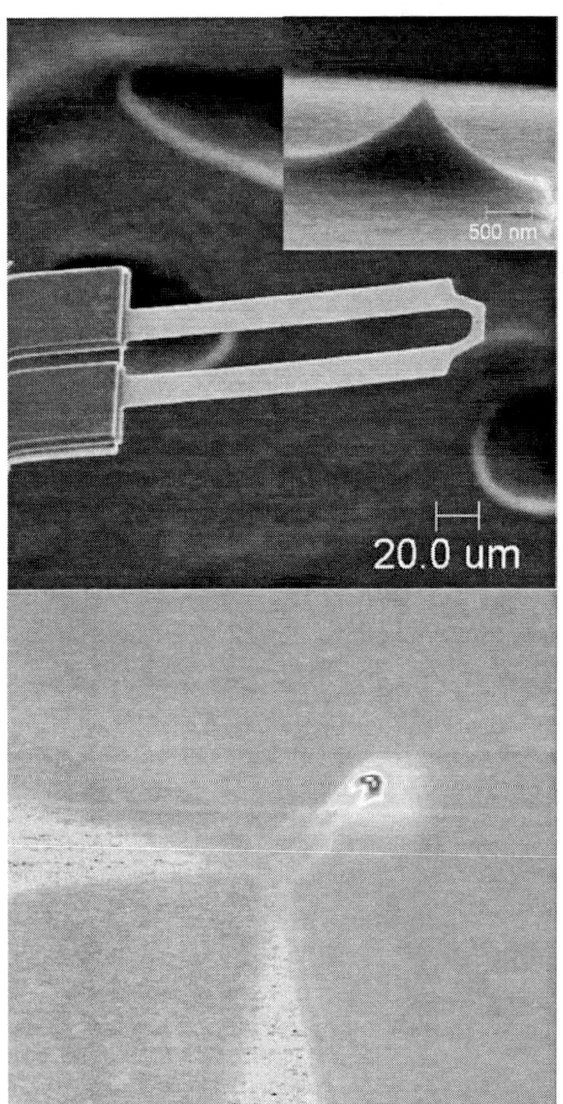

Figure 1: AFM cantilevers with integrated heaters fabricated at Georgia Institute of Technology. Top: Scanning electron microscope images. This cantilever has a tip with radius of curvature < 50 nm. Bottom: infrared microscope image of a cantilever during steady-state heating.

NANO2004-46075

Self-Assembled Silicon Microdevices Driven by Muscle

Jianzhong Xi, Jacob Schmidt, and Carlo Montemagno

Department of Bioengineering, University of California, Los Angeles,

7523 Boelter Hall, 420 Westwood Plaza, CA, 90095 Email: jzxi@ucla.edu; or cdm@seas.ucla.edu

Over the last two decades, a variety of micro-robotic systems have been developed including electrothermal, electrostatic, electrochemical, piezoelectric, and electromagnetic actuators based on MEMS technology. The development of these micro-actuators promises a revolution in biological and medical research and applications analogous to that brought about by the miniaturization of electrical devices in information technology. For example, controllable manipulation of these tiny actuators may enable precise temporal and spatial delivery of chemicals, micro-optics or microelectronics to specific targeted sites.

There has been much recent activity directed toward engineering devices powered by biological structures from the molecular to the tissue level. Since individual molecular motors provide only miniscule amounts of work, the actions of millions or more must be harnessed in parallel to result in significant activity in the macroscopic world. The prospects of exploiting natural massively parallel motor assemblies, such as muscle bundles, are very attractive since the organization, production, and manipulation of the motors from nanometer to millimeter length scales are coordinated by complex molecular machinery refined over millions of years of natural selection. Successfully engineering the muscle cells on current electric microchips is an initial but key step to fabricating autonomous intelligent hybrid micro-machines, which can be directly powered by glucose in the physiological fluids and respond to environmental stimuli.

Numerous cell patterning methods have been developed by using conventional and soft photolithography techniques[1, 2], and most of them are primarily suitable for patterning static cells on an immobile surface. However, the unique anisotropic, structural, and contractile physiological characteristics of muscle bundles complicate the construction of muscle-powered MEMS structures. Although several recent reports described the integration of the contractile cells on micropatterned elastic substrates or cantilevers, all studies focused on the sub-cellular level contraction and did not permit free motion of the cells and their integrated structures[3-6]. Therefore, a novel system of patterning the contractile cells spatially and selectively must be developed for the construction of the mobile hybrid structures.

Through manipulating material interfaces and phase transition, we have created a novel system for self-assembling muscle cells on MEMS devices. This system enables muscle tissues to self-assemble from individual cells at desired locations, remain tightly attached where desired, and allow the assembled muscle and its integrated devices freedom to contract. With these fabrication techniques, a group of microfabricated devices have been created. Self-assembling muscle cells on a cantilever has allowed us to perform in situ studies of mechanical properties of muscle cells. For example, Young's modulus and normalized maximum force of the rat neonatal ventricular myocytes are 40 kPa and 17.1mN/mm^2, respectively.

Engineering muscles to fabricate artificial robots, such as swimming fish, has been previously performed. However, unlike these conventional fabrication techniques, we addressed the self-assembly process, which eliminated the muscle bundle isolation step and the subsequent manual integration of the isolated tissues with mechanical devices. More importantly, in the context of our work, self-assembly refers to the ability of the muscle tissue bundle to automatically locate where desired and mature into a differentiated structure with a patterned anisotropy, resulting in the formation of a functional component, such as the legs described as below, without any manual assistance. In an engineering view, self-assembly is a dynamic and automatic process, which includes the growth and fusion of individual cells into muscle bundles, the integration of mature muscle bundles with microdevices, and the controlled release of them to freely contract.

Here we also present the first self-assembled muscle-powered micromachines (figure 1). The hybrid micromachines with the size of 160µm can walk for more than 4 hours, with maximum speeds of ~40µm/s, and such micromachines have the following advantages over conventional inorganic robots: autonomous (non-tethered), simple fabrication, and biologically friendly operating conditions.

In our work, we addressed several issues including the material interface, material compatibility, cell biology and nanofabrication, and have devised the strategy of self-assembly, which enable the integration and controlled release of muscle on microstructures. Since the fabrication of microstructures is totally independent from the assembly of cells, varieties of MEMS devices previously fabricated may be incorporated in our established system. This technology provides opportunities for development of a new class of mechanical/electrical bioactuators that can be powered directly from glucose present in physiological fluids.

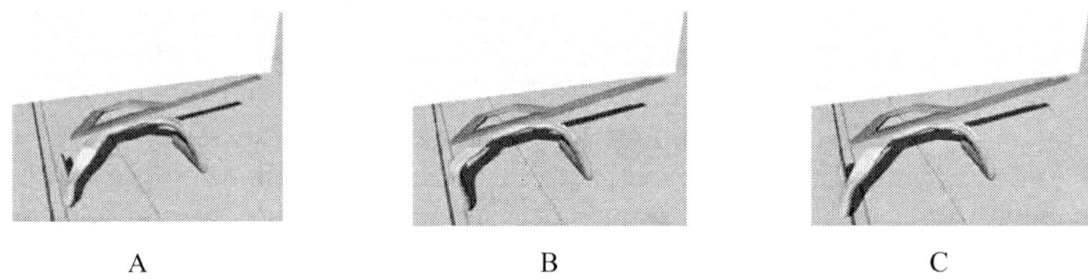

Figure 1. Schematic diagrams showing the sequential movement of the second microrobot during one step. A) before contraction of the leg. B) during contraction of the leg. C) after relaxation of the leg. Blue, green and red bars mark the start position of the inorganic scaffold and the motile leg, and the final position of the motile leg, respectively.

References:

[1]   A. Folch and M. Toner, "Microengeering of cellular interactions," *Annu. Rev. Biomed. Eng.*, vol. 02, pp. 227-56, 2000.

[2]   R. S. Kane, S. Takayama, E. Ostuni, D. E. Ingber, and G. M. Whitesides, "Patterning proteins and cells using soft lithography," *Biomaterials*, vol. 20, pp. 2363-2376, 1999.

[3]   N. Balaban, U. Schwarz, D. Riveline, P. Goichberg, G. Tzur, I. Sabanay, D. Mahalu, S. Safrant, A. Bershadsky, L. Addadi, and B. Geiger, "Force and focal adhesion assembly: a close relationship studied using elastic micropatterned substrates," *Nature Cell Biology*, vol. 3, pp. 466-473, 2001.

[4]   K. Beningo, M. Dembo, I. Kaverina, J. V. Small, and Y. L. Wang, "Nascent focal adhesion are responsible for the generation of strong propulsive forces in migrating fibroblasts," *Journal of Cell Biology*, vol. 153, pp. 881-887, 2001.

[5]   C. Galbraith and M. Sheetz, "A micromachined device provides a new bend on fibroblast traction forces," *Proc. Natl. Acad. Sci, USA*, vol. 94, pp. 9114-9118, 1997.

[6]   J. L. Tan, J. Tien, D. M. Pirone, D. S. Gray, K. Bhadriraju, and C. Chen, "Cells lying on a bed of microneedles: an approach to isolate mechanical force," *PNAS*, vol. 100, pp 1484-1489, 2003.

Proceedings of NANO2004
Integrated Nanosystems: Design, Synthesis and Applications
September 22-24, 2004, Pasadena, California, USA

NANO2004-46098

Mo$_2$C NANOWIRES AND RIBBONS ON Si VIA TWO-STEP VAPOR PHASE GROWTH

Loucas Tsakalakos
GE Global Research
Niskayuna, NY 12309
Email: tsakalakos@research.ge.com

Lauraine Denault
GE Global Research
Niskayuna, NY 12309

Michael Larsen
GE Global Research
Niskayuna, NY 12309

Mohamed Rahmane
GE Global Research
Niskayuna, NY 12309

Yan Gao
GE Global Research
Niskayuna, NY 12309

Joleyn Balch
GE Global Research
Niskayuna, NY 12309

Paul Wilson
GE Global Research
Niskayuna, NY 12309

EXTENDED ABSTRACT

Keywords: transition metal carbides, nanowires, synthesis, Si substrate

Transition metal carbides are an interesting class of electronic materials owing to their high electrical conductivity at room temperature, which is only slightly lower than that of their constituent transition metal elements. For example, the room temperature electrical resistivity of bulk Mo$_2$C is ~70 $\mu\Omega$-cm compared to that of Mo (4.85 $\mu\Omega$-cm), whereas that of NbC is ~50 $\mu\Omega$-cm as compared to 15.2 $\mu\Omega$-cm for Nb. Indeed, the temperature dependent resistivity of many transition metal carbides suggests metallic-like conduction. Furthermore, certain transition metal carbides are known to become superconducting, with transition temperatures ranging from 1.15 °K for TiC$_{1-x}$ to 14 °K for NbC. [1] They are also able to withstand high temperatures and are chemically stable. Initial synthesis of metal carbide nanorods was demonstrated using the carbon nanotube (CNT) confined reaction mechanism by Lieber and co-workers [2] and subsequent superconducting behavior was shown by Fukunaga *et al.* [3]. Vapor-liquid-solid growth was employed by Johnsson *et al.* [4] to synthesize micron-sized carbide whiskers. Here, we have successfully synthesized Mo$_2$C nanorods and ribbons on Si substrates using a novel two-step catalytic approach, which allows for synthesis of such high temperature nanostructures at manufacturable temperatures ($\leq$ 1000 °C) and time scales ($\leq$ 60 min). In the first step we utilize a catalytic vapor phase process to grow Mo and/or molybdenum oxide nanostructures, which are subsequently carburized *in situ* to form the desired Mo$_2$C nanostructures. Unlike true VLS growth of carbides, in which high temperature ($\leq$ 1100-1200 °C) is required to adequately dissolve carbon into the catalyst particles, our strategy is to react the nanostructures along their entire length with a carbon vapor source after creating the oxide/metal nanostructures, which for Mo$_2$C can be achieved at relatively low temperatures. ($\leq$ 1000 °C). The nanorods and ribbons are polycrystalline, with a mean grain size of 20-50 nm and 50-150 nm, respectively. We hypothesize that the growth mechanism is a complex mixture of VLS, VSS, and auto-catalytic growth, in which molten catalyst nanoparticles enter a three phase region once the metal precursor is supplied. The growth then presumably continues via a vapor-solid-solid process and is possible assisted by the presence of various molybdenum oxide species on the surface. Initial single nanowire electrical measurements yield a higher resistivity than in the bulk, which is attributed to the fine grain sizes and/or the presence of an oxide layer. A discussion of the growth mechanism will be presented along with issues relating to single nanowire device fabrication and control of nanowire orientation.

ACKNOWLEDGMENTS

The authors would like to thank the National Institute of Standard and Technology for financial support. We also thank Dr. M.L. Blohm and Dr. V. Mani for helpful discussions and support.

REFERENCES

[1] T. Ya. Kosolapova, <u>Carbides: Properties, Production, and Applications</u>, Plenum Press, New York, 1971.

[2] C.M. Lieber, E.W. Wong, H.-J. Dai, B.W. Maynor, L.D. Burns, "Growth and Structure of Carbide Nanorods," Mater. Res. Soc. Symp. Proc. **410**, 103 (1996).

[3] A. Fukunaga, S. Chu, and M.E. McHenry, "Synthesis, structure, and superconducting properties of tantalum carbide nanorods and nanoparticles," J. Mater. Res. **13**, 2465 (1998)

[4] M. Johnsson and M. Nygren, "Carbothermal synthesis of TaC whiskers via a vapor-liquid-solid growth mechanism," J. Mater. Res. **12**, 2419 (1997)

NANO2004-46024

INFORMATION MODELS AND VIRTUAL PROTOTYPING IN NANOMANUFACTURING

Dr. J. Cecil
Virtual Enterprise Engineering Laboratory (VEEL) Director,
Center for Information Based Manufacturing
Department of Industrial Engineering
New Mexico State University
Las Cruces, NM 88003
jcecil@nmsu.edu

J. Senthil Nathan, N. Gobinath
Graduate Assistants
Center for Information Based Manufacturing
New Mexico State University

ABSTRACT

This paper discusses the use and creation of virtual reality based models and information oriented process models related to the study of Nanomanufacturing. EML, (which is an evolving language), was originally proposed [1] to enable the capture of information rich contexts at various levels of the manufacturing levels. In this paper, the use of EML models as a basis for describing and analyzing nano manufacturing processes will be described; their subsequent use in the creation of simulation models is also discussed. A framework for analysis, reasoning and simulation is proposed which includes a virtual reality based simulation module. Our experience in using VRML 2.0 (Virtual Reality Modeling Language) for the creation of such virtual reality environments will also be addressed.

INTRODUCTION

There are two foci in this paper: (1) The usefulness of information models and information modeling languages such as the Enterprise (Engineering) Modeling Language (EML) as a basis for planning, analysis and mapping target nanoscale manufacturing processes (2) Research issues related to the creation of virtual reality based simulation environments, in which nanoprocesses are modeled.

The paper will conclude on the role of the above two approaches / principles in teaching engineering students fundamental concepts in nanoscale manufacturing. Findings from a NSF funded project will also be discussed.

The long-term objectives of this joint research and educational initiative are to explore the design of a Virtual Reality based simulation framework for nano processes. As the study of nanoscale processes is still a research oriented activity, we are designing a simulation test bed whose capabilities and scope of application will continue to expand based on additional findings and approaches from the progress of our research efforts and other initiatives.

ROLE OF INFORMATION MODELS

The creation of an 'information oriented' or 'information intensive' model for a target set of nanoscale activities is part of the overall approach, which seeks to characterize in detail various attributes related to a target nanoscale process. The information models under development are based on an evolving language called Enterprise Modeling Language (EML), which has proven to be useful in modeling micro systems and processes. For the nanoscale processes, the creation of such an information model will help identify several categories of information such as influencing criteria, associated performing agents, as well as the final or intermediary outcomes. These models can be built at various levels of abstraction; such decompositions can capture the relationships among various sub-functions or sub-tasks using attributes termed as end effectors.

For a target nanoscale process (for example, assembly of nano particles using an Atomic Force Microscope probe as a gripper), a formal information model would be useful by capturing the salient attributes, which influence that process. Influencing criteria including constraints (such as the process and equipment capabilities relating to the experimental setup, schedule, etc), information inputs (that are needed to begin a nanoscale process), and physical inputs (such as the materials, catalysts and other chemicals) can be modeled explicitly. The associated performing agents (which are physically involved in a target nanoscale process) can be categorized into physical agents (such as the lab personnel, etc.), software and physical resources. The physical resources refer to the equipment, tools and other devices used in the laboratory associated with accomplishing a nanoscale process. Using this approach, detailed models of nanoscale processes can be built. These models can facilitate better understanding of the complex relationships among the various entities and agents in a given processes as well as provide a formal basis for 'driving' the animation or simulation of a specific nanoscale process.

FRAMEWORK FOR A SIMULATION BASED TEST BED

The proposed framework for a simulation test bed includes the following components (figure 1): an information model, a knowledge base, a simulation module and an immersive equipment module.

The information model (for a target nanoscale process or set of processes) drives the simulation activities. This model is created using an Enterprise Modeling Language (EML) [1].

The knowledge base contains rules, heuristics, and other information related to explaining the occurrence of various steps, phases or activities within that target nanoscale process or processes; for a given context such as an assembly process (for example), the knowledge base attempts to propose process steps as hypotheses, whose occurrence will result in a target product. The reasoning about assembly or process alternatives depends on the user inputs and the rules / heuristics in the knowledge base.

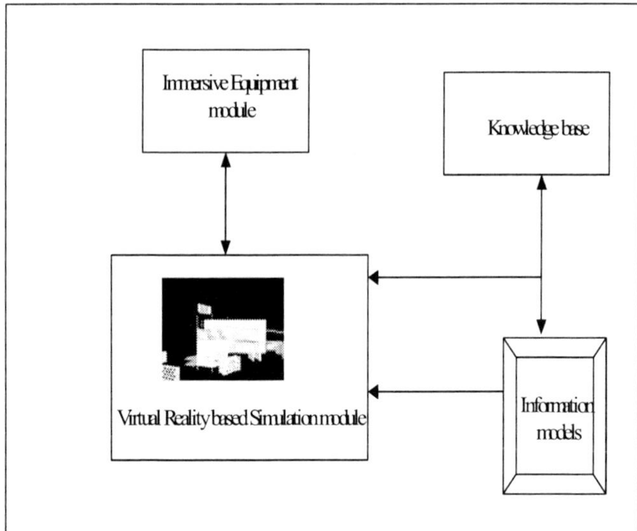

Figure 1: Components of the Simulation Test bed for nanoscale processes

The third component in this framework is a Virtual Reality based simulation module (which comprises of graphical processors, parsers, data bases, planning resources for assembly and/or path planning pertaining to the nanoscale processes, etc.). The overall simulation activities are controlled by the

process map or sequence contained in the associated EML model. The events to be simulated are stored in the EML model, whose attributes are populated with the help of the outcomes generated by the Knowledge Base described earlier.

The last component is an Immersive Equipment module; this module manages the interaction of users with the Virtual Reality (VR) environment through devices such as Wands, 3 D mouse, stereo eyewear, etc. With the help of this interface, users can immerse themselves in the VR environment.

The overall approach involves an EML based information model driving the various VR simulations. For a given set of conditions, the objective of the simulation test bed is to propose process or assembly outcomes with the help of a knowledge base (which comprises of facts and rules); a candidate process alternative or a set of process alternatives is used to populate the various information attributes in its corresponding EML model. For example, the needed material inputs and other information drivers can be captured in the EML model; in a similar manner, the constraints influencing the process can be captured as well (which could range from temperature / experimental conditions, resources, presence of catalysts, etc.) along with the performing agents (which refers to the physical resources involved in the target process under reference).

This paper [2] highlights our preliminary work relating to the design and development of various information models as well as the design of the Virtual Reality based simulation environment. Our experience in using a language called Virtual Reality Modeling Language 2.0 (VRML) to model nanotubes and visualizes nanoscale assembly scenarios will also be addressed. The information models and virtual environments can be used to educate undergraduate and graduate students on nanoscale processes and systems as well. A summary of the potential challenges involved in the implementation [3] of the simulation test bed will also be provided.

ACKNOWLEDGMENT

Funding for the research and educational activities described in this paper have been received from the National Science Foundation (grant number 0304269).

REFERENCES

1. Xavier, B., Cecil, J. (2001). Design of an Enterprise Modeling Language (EML). Technical Report. Virtual Enterprise Technologies, Inc. (VETI). Las Cruces, New Mexico.

2. VEEL Website: http://web.nmsu.edu/~jcecil/VEEL.html (Under section Research & Education, Nanotechnology Research)

3. Cecil, J., NUE Nanotechnology Project Report, Center for Information Based Manufacturing, Industrial Engineering Department, New Mexico State University, Las Cruces, NM 88011.

NANO2004-46043

C-MEMS/NEMS: A Novel Technology for Nanoscale Material Formation from Graphite Fiber to Ni and Si Nanowires

Chunlei Wang, Rabih Zaouk, Kartikeya Malladi, LiliTaherabadi, Marc Madou[1]
Department of Mechanical and Aerospace Engineering, University of California, Irvine
Irvine, CA92612

Carbon microelectromechanical systems (C-MEMS) and carbon nanoelectromechanical system (C-NEMS) have received much attention because of the many potential applications. Some important applications include: DNA arrays, glucose sensors, microbatteries and biofuel cells. Microfabrication of carbon structures using current processing technology, including focused ion beam (FIB)[1] and reactive ion etching (RIE)[2], is time consuming and expensive. Low feature resolution, and poor repeatability of the carbon composition as well as widely varying properties of the resulting devices limits the use of screen printing of commercial carbon inks for C-MEMS. Our newly developed C-MEMS microfabrication technique is based on the pyrolysis of photo patterned resists[34]. Figure 1(a) shows a typical SEM image of C-MEMS/NEMS features with carbon posts connected by carbon fibers. Figure 1(b) shows a typical carbon post with carbon nanofibers on its side surfaces.

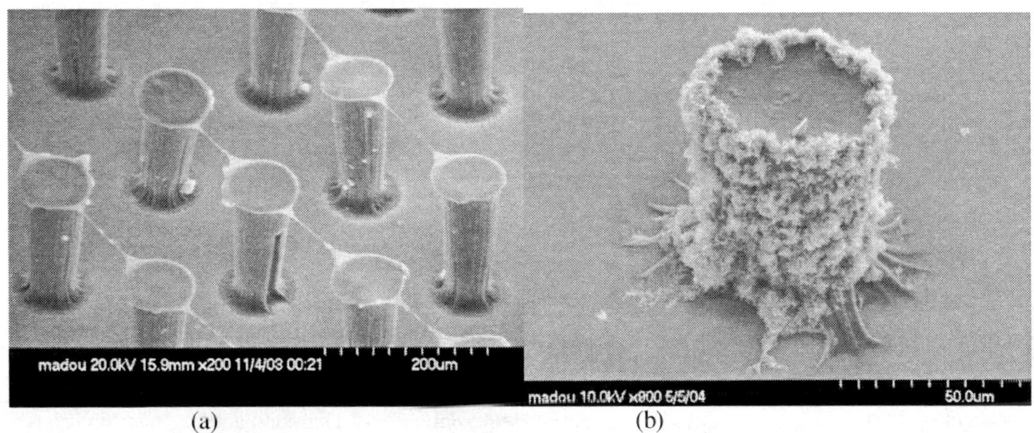

Figure 1. typical SEM pictures of (a) C-MEMS/NEMS structure with carbon fibers connected by carbon posts, and (b) a carbon post with nanofibers grown around the side walls.

Unlike conventional CVD methods for growing carbon fibrous materials and carbon nanotubes in which a gaseous carbon source, such as CH_4, is commonly used, we use photoresist as the carbon source. Using suitable catalysts, graphite nanofibers and Ni nanowires can be formed as well. Furthermore, Si nanowires were successfully grown without photoresist patterns and a modified solid-liquid-solid (SLS) mechanism was used to explain our results.[5] Figure 2 shows TEM images of a single crystalline Ni nanowire. Detailed SEM and TEM investigations were performed.

---

[1] Email: mmadou@uci.edu

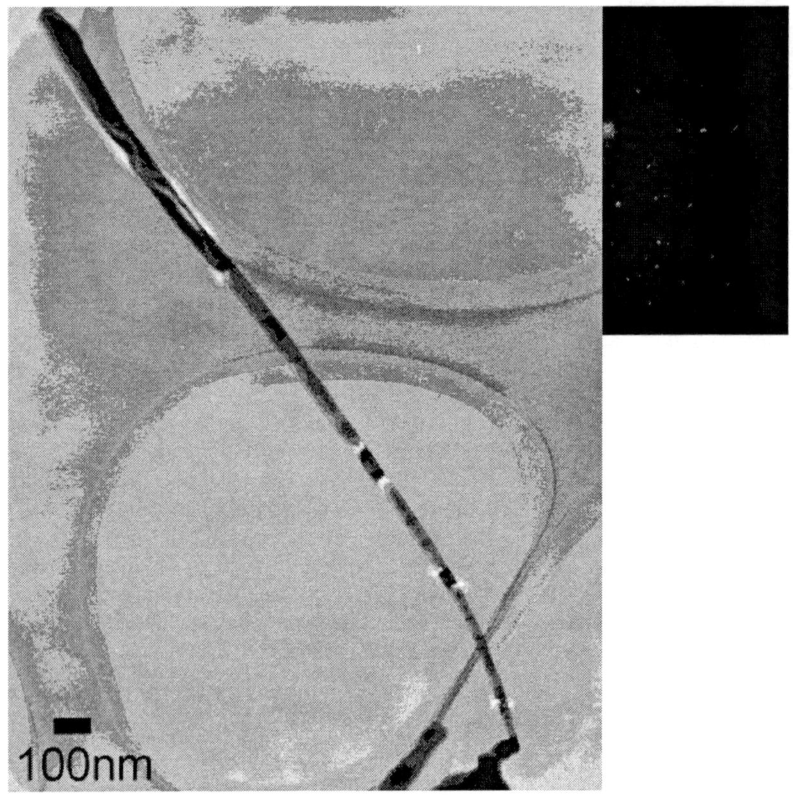

Figure 2. A Typical TEM image of a Ni nanowire and its diffraction pattern.

[1] Miura, N., Numaguchi, T., Yamada, A., Konagai, M., Shirakashi, J., 1998. Room temperature operation of amorphous carbon-based single-electron transistors fabricated by beam-induced deposition techniques. Japanese Journal of Applied Physics, Part 2: Letters, vol. 37, L423-L425.

[2] Irie, M., Endo, S., Wang, C.L., Ito, T., 2003. Fabrication and properties of lateral p-i-p structures using single crystalline CVD diamond layers for high electric field applications. Diamond and Related Materials, 12, 1563-1568.

[3] Kim, J, Song, X., Kinoshita, K., Madou, M.and White, R., 1998. Electrochemical studies of carbon films from pyrolyzed photoresist. J.Electrochem.Soc., 145, 2314-2319.

[4] Ranganathan, S., McCreey, R., Majji, S. M., and Madou, M., 2000. Photoresist-derived carbon for microelectromechanical systems and electrochemical applications, J.Electrochem.Soc., 147, 277-282.

[5] C.L.Wang, M.Madou et al, to be submitted

NANO2004-46062

SURFACE-WAVE-SCATTERING-BASED CHRACTERIZATION OF NANO-PARTICLES

Mustafa M. Aslan and M. Pinar Mengüç

Mechanical Engineering, University of Kentucky
Ralph G. Anderson Building, Lexington, KY, 40506, USA

ABSTRACT

In this paper, we introduce a novel methodology to determine the size and structure of nano-particles on a surface. We present an analysis, dubbed Elliptically Polarized Surface-Wave Scattering (EPSWS) approach, to show that 5-10 nm size particles on or above an interface can be characterized by using the scattered surface plasmon (SP) or evanescent waves. We present an analysis to show that the scattering matrix elements of the evanescent waves scattered by the nano-particles on or near an interface can be used for characterization of nano-size particles.

INTRODUCTION

Nano-particles (1-100 nanometer in effective diameter) are known for their unusual properties that are due mostly to their small sizes and high percentage of atoms in surface states, which yield unique properties that differ from those of the same bulk materials [1]. Using metallic nano-particles or colloids, it is possible to obtain unprecedented optical, electrical, and structural properties if the composition, structure, shape, and size distribution of nano-building-blocks can be controlled during fabrication. Yet, to be able to engineer bottom-up processes at nanoscale, new approaches need to be developed for measurement and visualization of these particles and structures. Furthermore, such measurements must be incorporated as real time diagnostic tools for reliable manufacturing applications.

Even though scattered light is used routinely for non-intrusive characterization and visualization of structures and surfaces in real time, it cannot be adapted readily for the identification of nano-structures. The wavelengths of typical light/laser sources available are usually hundreds of times larger than the size of the nano-particles that need to be characterized. Consequently, traditional light scattering/ absorption techniques cannot infer the detailed and accurate size and structural information of such small particles.

In this paper, we introduce a novel methodology to determine the size and structure of nano-particles. We present an analysis, dubbed Elliptically Polarized Surface-Wave Scattering (EPSWS) approach, to show that 5-10 nm size particles on or above an interface can be characterized by using the scattered surface plasmon (SP) or evanescent waves. The evanescent waves and surface plasmons are the result of total internal reflection that takes place at the opposite side of the medium from incident light [2]. If there are any particles within the reach of the surface waves, the energy is tunneled to the particle and then scattered. An experimental study by Sterligov *et al.* [3] indicated that the nano-size particles on a surface can scatter evanescent waves and may cause a change in the polarization of the incident light. However, no potential characterization methodology was offered. We present an analysis of the problem, based on the theoretical formulation reported in [4], to show that the scattering matrix elements of the evanescent waves scattered by the nano-particles on or near an interface can be used for characterization of such particles. In several previous studies, we have shown that characterization of small particles and agglomerates can be described efficiently through polarized light scattering [5,6]. Four normalized scattering matrix elements, M_{11}, M_{12}, M_{33} and M_{34} are particularly important for this purpose. In the present study, we show for the first time that these patterns are also crucial for characterization of nano-size particles based on scattering patterns of evanescent-waves.

ELLIPTICALLY-POLARIZED-SURFACE-WAVE-SCATTERING MODEL

Surface plasmon (SP) waves are the quanta associated with longitudinal waves propagating in matter through the collective motion of large numbers of electrons. As in EWs, the intensity of surface EM radiation (ISW) has a maximum at $z = 0$, $x = 0$ and exponentially decreases along the z-axis. The fundamentals of SP waves are well studied in the literature and relatively well understood [2]. SP response of a medium is generally very sensitive to the complex refractive index of the medium. SP waves have been used extensively in the development of optical sensors for measurement of chemical and biological components [7], and characterization of surface and thin films [8].

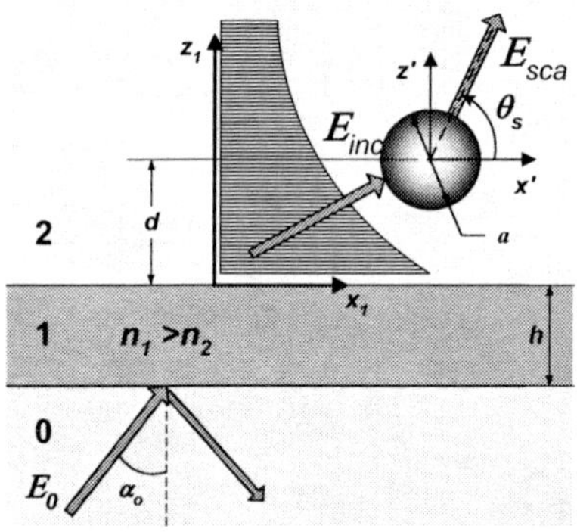

Fig.1: Geometry of a sphere near/on a metallic thin film. Medium 0, 1, and 2 are glass, thin gold film and suspension (ethyl alcohol with metal sphere), respectively.

To our knowledge, there is no published research in the open literature about the potential use of polarized surface waves for characterization of nano-particles and nano-structures on a metallic thin film, other than a few experimental approaches. Natan's group studied SP and Au particles experimentally [8]. Their approach, however, is based on the absorption by a thin film, and does not take into account the elliptically polarized scattering signatures of particles for size determination. Sterligov *et al.* [3] investigated the particle-surface wave interaction and carried out both the near- and far-field measurements. However, they did not suggest any approach to characterize nano-particles *in situ*. In the present study, we show that the use of Mueller matrix elements allows characterization of nano-particles using the elliptically polarized light-scattering patterns.

The geometry of the scattering system considered in this study is shown in Figure 1. The incident radiation is a plane wave whose wave vector is in the *x-z* plane, oriented at angle α_0 with respect to the *z* axis, traveling from Medium 0 (m-0; bottom) to m-1 (middle). This wave undergoes total internal reflection at the interface between the m-0 and m-1, which results only in standing surface waves in m-1 and m-2. Medium 1(m-1) is a thin metallic film with thickness h and m-2 is an ethyl-alcohol suspension of metallic nano-particles. For the general case, we consider a particle located a distance d above the interface separating m-1 from m-2. The wave incident on the particle is assumed to be parallel to the interface. The general formulation considered in [4] allows any orientation of the planar wave incident on the particle and accounts for particle-surface interactions. Particles are assumed to convert (tunnel) the standing evanescent wave to the particle and after that to propagate scattering waves, as if they are probes.

Stokes vector of a scattered light which contains the flux and polarization information of the medium can be related to the incident light Stokes vector at given wavelength scattering amplitude matrix. This relationship can be written in matrix form for a spherical particle in a symmetric medium [9]. The scattering matrix elements (S_{11}, S_{12}, S_{33} and S_{34}) can be calculated from the scattering amplitudes (S_1, S_2, S_3, and S_4) using the equations outlined in [9]. In order to explain scattering matrix elements' results better, we define a normalized scattering matrix and its elements are $M_{11}=S_{11}$, $M_{12}=S_{12}/S_{11}$, $M_{33}=S_{33}/S_{11}$, $M_{34}=S_{34}/S_{11}$. In this paper we represent normalized scattering matrix elements of spherical nanoparticle as M_{11}, M_{12}, M_{33}, and M_{34}.

RESULTS AND CONCLUSIONS

In numerical experiments, we consider a thin gold film (Medium 1, m-1) on a glass substrate (m-0). On top of the gold film is an ethyl-alcohol solution (m-2) that may contain a suspension of metallic spherical nano-particles. The value of the EM-wave amplitude on the particle depends on several parameters, including the film thickness, refractive indices of m-1 and m-2, as well as the incident angle α_0. It is important to note that the material properties of nano-particles alter the overall effective properties of the m-2. Therefore, depending on the nano-particles' material properties and the volume fraction, the amplitude of the evanescent waves in m-2 also vary. This amplitude is important as the amount of energy scattered by a particle depends on its value. The particle volume-fraction effects on the effective refractive index of the medium may be considered via Maxwell-Garnett theory. Both the particle volume fraction and particle refractive index alter the effective index of refraction of the medium, which change the results. The values we use here are representative and only for the purpose of parametric study; they need to be modified according to the conditions dictated by actual experiments.

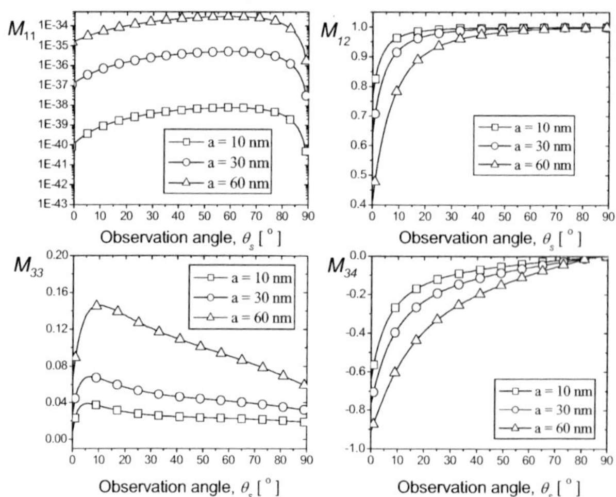

Fig.2: M_{11}, M_{12}, M_{33} and M_{34} profiles of the light scattered by gold nanoparticles on the surface of the gold film ($d - a/2 = 0$ nm).

We calculate the angular profiles of the scattering matrix elements following the formulation outlined in [4]. The effect of the diameter of a spherical gold nano-particles on M_{11}, M_{12}, M_{33} and M_{34} profiles as a function of observation angle is shown in Fig. 2. Here, it is assumed that the nano-particle is resting on the gold-film. These results show that M_{11} profiles peak around $\theta_s = 55°$ and increases with particle size, as the larger particles intersect more of the incident field flux. All three polarization matrix elements show differences with particle size that may be used for detailed characterization.

Copyright © 2004 by ASME

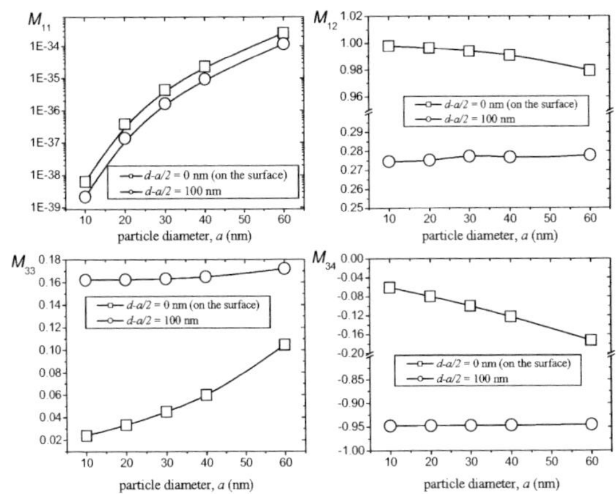

Fig.3: Profiles of M_{ij} for scattered light by gold nanospheres located on a thin gold film ($d -a/2= 0$), and $d-a/2 = 100$ nm; observation angle is $45°$, film thickness is 50 nm, incident angle for the wave is $70°$.

The sensitivity of predictions to particle location is depicted in Figure 3 from a different perspective, where only the results for scattering direction of $45°$ are shown. Note that these are representative results for a single observation angle; it is possible to combine the results from multiple observation angles to have more robust characterization. It is obvious that M_{11}, M_{12}, M_{33}, and M_{34} profiles can be used together to identify the particle sizes accurately on the surface of the film. The results suggest that it is possible to identify particle sizes between 10 and 60 nm. On the other hand, if particles are at about 100 nm above the surface, the characterization may not be possible. The combination of determining both particle size and surface separation distances may pose a challenge; however, these cases need to be examined in conjunction with careful experimentation, which is currently being conducted.

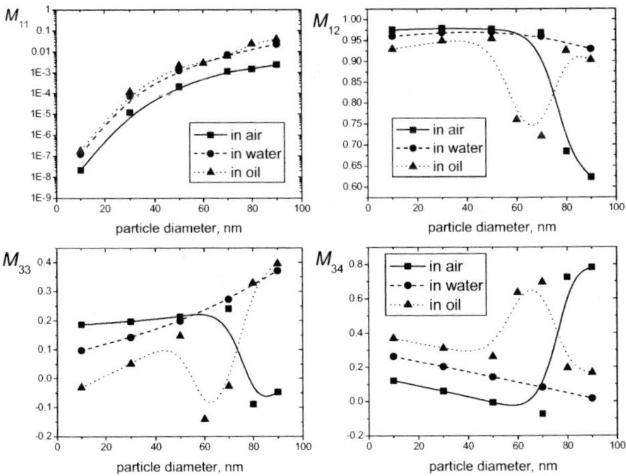

Fig.4: M_{ij} for scattered light by gold nanospheres located on a thin gold film ($d -a/2= 0$) in three different medium (air, water and oil); observation angle is $45°$, film thickness is 50 nm, incident angle for the wave is $70°$.

Figure 4 shows how the type of the medium that surrounds the particles affects the scattering results. M_{ij} for gold nano-particles in three different types of medium; air, water and oil were plotted at 45 scattering angle. Refractive index of air, water and oil are assumed to be 1, 1.33, and 1.5, respectively. Although M_{12} results are not very sensitive to particle size in 10-100 nm range, M_{33} shows significant difference when size of particles changes from 10 nm to 100 nm. Water is the best solution if the gold nano-particles need to be characterized. Further detailed work needs to be carried out before applying these results to actual problems.

CONCLUSIONS

Based on the results presented here, metallic, spherically shaped nano-particles on a metallic thin film can be characterized from the normalized scattering matrix elements M_{ij}. These results are encouraging, as there are no other techniques available to characterize metallic nano-particles reliably. It is important to extend these results to different size distributions as well as to the agglomerates of metallic particles (colloids) and evaluate these findings against the experiments.

ACKNOWLEDGMENTS

This work is sponsored by Kentucky Science and Education Foundation, University of Kentucky Office of Vice President, and by a NSF-NER (2004) grant

REFERENCES

[1] Feldheim, D.L., and Foss, C.A., *Metal Nanoparticles: Synthesis, Characterization, and Application*, Marcel Dekker Inc., New York , 2002.

[2] Raether, H., *Surface Plasmons on Smooth and Rough Surfaces and on Gratings*, Springer-Verlag, Berlin, 1988.

[3] Sterligov, V.A., Cheyssac, P., Lysenko, S.I., and Kofman, R., Surface plasmon-polaritons II. Scattering by metallic nanoparticles, *Physica Status Solidi A-Applied Research* Vol.175, pp 259-264, 1999.

[4] Videen, G., Aslan, M. M, and Mengüç, M.P., Characterization of Metallic Nano-particles Via Surface Wave Scattering: A. Theoretical Framework and Formulation, *International Symposium on Radiative Transfer*, June 20-25, 2004, Istanbul, Turkey.

[5] Manickavasagam, S., Mengüç, M. P., Drozdowicz, Z. B.and Ball, C., Size, Shape, and Structure Analysis of Fine Particles, *American Ceramic Society Bulletin*, Vol.81, No.7, pp.29-33 July 2002.

[6] Aslan, M., Yamada, J., Mengüç, M. P., Thomasson, A., "Characterization of Individual Cotton Fibers via Light Scattering: Experiments," *AIAA Journal of Thermophysics and Heat Transfer* Vol. 17, No. 4, pp. 442-449, 2003.

[7] Homola, J, Yee, S.S., and Gauglitz, G., Surface Plasmon Resonance Sensors: Review, *Sensors and Actuators B-Chemical*, Vol. 54, No.1-2, pp 3-15, 1999.

[8] Lyon, L.A., Musick, M.D., and Natan, M.J., Colloidal Au-enhanced surface plasmon resonance immunosensing, *Analytical Chemistry*, Vol.70, pp. 5177-5183, 1998.

[9] Bohren, C.F., and Huffmann, D.R., *Absorption and Scattering of Light by Small Particles*, Wiley, New York, 1983.

NANO2004-46087

ANALYSIS OF FLUID FLOW IN CYLINDRICAL MICROCHANNELS SUBJECTED TO UNIFORM WALL INJECTION

Mohammad Layeghi
Ph. D. (5) student, School of Mechanical Engineering
Sharif University of Technology, P.O.Box: 11365 – 9567, Tehran, Iran
Tel:+98216165685, Fax:+98216000021
Email: mlayeghi@mehr.sharif.ir

ABSTRACT

Analytical analysis of fluid flow in cylindrical microchannels subjected to uniform wall injection at various Reynolds numbers is presented. The classical Navier-Stokes equations are used in the present study. Mathematically, using an appropriate change of variable, Navier-Stokes equations are transformed to a set of nonlinear ordinary differential equations. The governing equations are solved analytically using series solution method. The presented analytical results can be used for the prediction of velocity profiles and pressure drops in the cylindrical micro channels. The results are validated against available data in the literature and have shown good agreement.

KEYWORDS: Laminar and Incompressible Fluid Flow, Cylindrical Microchannel, Navier-Stokes Equations, Analytical Method.

INTRODUCTION

The ability to analyze fluid flow in microchannels has great importance in designing and utilization of microfluidic devices. Generally, microfluidics refers to devices or flow configurations that have the smallest design feature on the scale of a micron or larger. Modern developments of microfluidic devices for fluid transport have found many applications, ranging from the life sciences industries for pharmaceuticals and biomedicine to industrial applications of combinatorial synthesis [1]. Other areas of applications for micro devices for the transport of liquids and gases include aerospace and automotive industries, printing, and optical applications. Novel electrical devices assembled using microfluidic components are also a possible area for technological innovation [2,3]. Micro devices are useful because they allow manipulation with fast response times; they can handle small fluid volumes, sense and control flows and pattern substrates on small length scales.

Typical applications are concerned with gases, water, or other aqueous solutions. The channels manufactured to date have dimensions between 1 to $300\,\mu m$, and flow speeds u, though varying widely, and at least for liquids they are possible in the range up to cm/s, which yields a Reynolds number $\mathrm{Re} = \rho u l / \mu < 30$ where ρ and μ denote the density and viscosity of the fluid. In many cases, although the Reynolds number is not actually less than 1, it is in such a range that the flow remains laminar.

Investigations of low-Reynolds number hydrodynamics naturally fall into the fluid dynamics realm of long interest to the chemical and mechanical engineering community [4,5]. In deed, in 1970s, Batchelor coined the term micro-hydrodynamics to describe this flow regime. The most important new themes introduced by the small length scales of microfluidic devices are the significant role of surface forces. Fluid motions in these small-scale systems are usually driven by applied pressure differences, electric fields, capillary forces, gradients in interfacial tension, etc [6-7].

Some discrepancies have been reported between flow measurements made in small channels and experiences from classical theory based on solutions to the Navier-Stokes equations. However, a careful consideration of the experimental results for pressure driven flows of liquids demonstrates that, in fact in almost all cases there are no significant discrepancies [8]. It is clearly important to recognize the significant influence geometry plays in low-Reynolds number flows, as the familiar poiseuille formula (or a lubrication calculation) makes clear. For example, since the pressure drop as a function of flow rate varies as the inverse fourth power of the radius, a small change in the dimension transverse to the flow, say due to manufacturing imperfections or objects adhering to the boundary, produces large changes in the flow.

For gases, the theory is in excellent agreement with experiment if compressibility and finite wall slip (owing to the mean free path being comparable to the channel dimensions) are properly accounted for [9]. Furthermore, it is worth recognizing that electrical effects due to charged double layers can be difficult to quantify, in particular since charge inhomogenities may occur. Also at the scale of hundreds of nanometers, at least for surfaces that are essentially atomically smooth, there is evidence for slip [10]. Thus it is our understanding that for small molecule liquids such as water, the familiar continuum description remains an appropriate starting point for analysis of micro devices, with appropriate consideration being given to electrical effects, slip, etc., as indicated briefly before.

Another approach for the study of fluid flow and transport properties in micro and nano channels is molecular dynamics method which is usually called molecular dynamics simulation or MD simulation. Molecular dynamics is widely used to simulate many particle systems ranging from solids, liquids, gases, and biomolecules on earth, to the motion of stars and galaxies in the universe. In a MD simulation, the motion of individual atoms within an assembly of N atoms or molecules is modeled on the basis of either a Newtonian deterministic dynamics or a Langevin-type stochastic dynamics, given the initial position coordinated and velocities of the atoms. There is an extensive literature on the MD simulation [11-14]. However, MD simulation has a number of limitations including time and length scale limitations, number of particles, and can not be used for a wide range of problems [15].

The basic idea of this paper is to present an innovative exact solution of Navier-Stokes equations which seems to be valid in wide ranges of dimensions including nano, micro, and higher scales and wide ranges of Reynolds numbers. The Navier-Stokes equations are solved for the study of fluid flow in a cylindrical microchannel subjected to uniform wall injection. The model results are also validated against available data in the literature. The presented analytical results can be employed for the analysis of vapor flow in micro heat pipes.

NOMENCLATURE

C_p	heat capacity at constant pressure
Kn	Knudsen number
L	injection length
$\dot{m}$	mass flow rate
Ma	Mach number
P	pressure
R	microchannel radius, gas constant
Re	Reynolds number
r	radial coordinate
T	temperature
V	injection velocity
v	radial velocity
w	axial velocity
z	axial coordinate

Greek Symbols

γ	specific heat ratio
μ	viscosity
ρ	density
ϑ	kinematic viscosity

Subscripts

f	fully developed
g	gas
sat	saturation
r	radial
z	axial

GOVERNING EQUATIONS

The axisymmetric motion of a fluid flow in a cylindrical microchannel subjected to uniform wall injection is considered here (Fig 1). The present model is based on the following assumptions:

(1) The flow is laminar and in steady-state condition and the fluid is incompressible;

(2) The properties of the fluid are assumed to be constant;

(3) Uniform injection occurs on the pipe wall from $z = 0$ to $z = L$ along the pipe (Fig 1);

(4) No-slip boundary condition is assumed on the microchannel wall;

(5) The left end of the microchannel ($z = 0$) is closed.

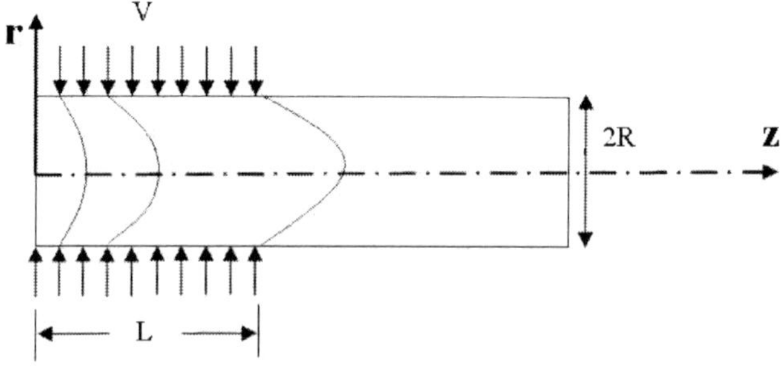

Fig.1 Fluid flow in a cylindrical microchannel subjected to uniform wall injection along with coordinate system

Copyright © 2004 by ASME

The fluid flow is described by the Navier-Stokes equations, together with the mass conservation equation. These equations are written in their conservative form as:

$$\frac{\partial w}{\partial z} + \frac{1}{r}\frac{\partial}{\partial r}[r\,v] = 0 \tag{1}$$

$$\rho\left(w\frac{\partial w}{\partial z} + v\frac{\partial w}{\partial r}\right) = -\frac{\partial P}{\partial z} + \mu\left[\frac{\partial^2 w}{\partial z^2} + \frac{1}{r}\frac{\partial}{\partial r}\left(r\frac{\partial w}{\partial r}\right)\right] \tag{2}$$

$$\rho\left(w\frac{\partial v}{\partial z} + v\frac{\partial v}{\partial r}\right) = -\frac{\partial P}{\partial r} + \mu\left[\frac{\partial^2 v}{\partial z^2} + \frac{1}{r}\frac{\partial}{\partial r}\left(r\frac{\partial v}{\partial r}\right) - \frac{v}{r^2}\right] \tag{3}$$

where w and v are the velocity components in the z and r directions, respectively and P is the fluid pressure. ρ and μ are the density and viscosity of the fluid, respectively. The boundary conditions are defined as follows.

$$w(0,r) = v(0,r) = 0, \quad \frac{\partial w}{\partial r}(z,0) = v(z,0) = 0,$$

$$w(z,R) = 0, \quad v(z,R) = V, \quad P(0,r) = 0 \tag{4a-e}$$

where L is the injection length on the microchannel, and R is the microchannel radius.

METHOD OF SOLUTION

The governing equations can be transformed to a set of non-linear ordinary differential equations using appropriate change of variables based on the physics of fluid flow in cylindrical pipes. Only one symmetrical half of the microchannel longitudinal section is considered in the present analysis. Previous numerical, and analytical investigations of the author and taking into account the boundary conditions (4a-e) show that the following assumptions can be made in addition to assumptions (1)-(5) in the previous section at various radial Reynolds numbers:

$$w(z,r) = z\,G(r), \quad v(z,r) = H(r) \tag{5a-b}$$

Equation (5a) can be easily deduced from the conservation of mass and boundary conditions for axial velocity. The mass flow rate at section z can be written as

$$\dot{m}(z) = 2\pi R \rho V z = \rho \pi R^2 \, \overline{w}(z) \tag{6}$$

where $\overline{w}(z)$ is the average flow velocity at section z in the microchannel. Rearranging Eq. (6) gives:

$$\overline{w}(z) = 2V\frac{z}{R}, \quad V > 0 \tag{7}$$

Assuming $w(z,r) = f(z)g(r)$ and using the definition of average velocity we obtain

$$\overline{w}(z) = \frac{\displaystyle\int_0^R f(z)g(r)\,2\pi r\,dr}{\pi R^2} = c_1\,f(z) \tag{8}$$

where c_1 is a constant value. Comparing Eqs. (7) and (8) yields $f(z) = c\,z$, where c is a constant value. Therefore, $w(z,r)$ can be assumed to have the form $w(z,r) = c\,z\,g(r)$. Finally, $w(z,r)$ should have the form like that expressed in Eq. (5a). Assumption (5b) can be easily deduced from Eq. (1) by nothing that $\partial w/\partial z$ is independent of z since $w(z,r)$ assumed to be in the form of Eq. (5a). Assumptions (5a,b) are valid in wide ranges of radial Reynolds numbers and for various microchannel diameters.

Substitution of Eqs. (5a,b) into Eqs. (1)-(3) gives the following set of ordinary differential equations:

$$G + \frac{1}{r}H + H' = 0 \tag{9a}$$

$$\frac{\partial P}{\partial z} = \left\{\mu\left[\frac{1}{r}G' + G''\right] - \rho\left[G^2 + H G'\right]\right\}z \tag{9b}$$

$$\frac{\partial P}{\partial r} = \mu\left(H'' + \frac{1}{r}H' - \frac{1}{r^2}H\right) - \rho H H' \tag{9c}$$

where ϑ is the kinematic viscosity of the fluid.

Differentiating from Eqs. (9b,c) with respect to r and z, respectively and combining them by removing the pressure gives

$$\frac{d}{dr}\left\{\mu\left[\frac{1}{r}G' + G''\right] - \rho\left[G^2 + H G'\right]\right\} = 0 \tag{10}$$

Substitution of function G from equation (9a) in terms of function H into equation (10) gives

$$\vartheta\left(r^4 H^{(iv)} + 2r^3 H''' - 3r^2 H'' + 3rH' - 3H\right)$$
$$= r^4 H''' H - r^3 H'' H - r^4 H'' H' - r^3 H'^2 - 3r^2 H' H + 4rH^2 \tag{11}$$

Equation (11) is a fourth order nonlinear differential equation which needs four boundary conditions to be solved. The associated boundary conditions in terms of $H(r)$ can be obtained from Eqs. (4b-d) which are:

$$H(r = 0) = 0, \quad H(r = R) = -V \tag{12a,b}$$

$$\left(\frac{1}{r}H + H'\right)_{r=R} = 0, \quad \left(-\frac{1}{r^2}H + \frac{1}{r}H' + H''\right)_{r=0} = 0 \tag{12c,d}$$

Equation (11) along with boundary conditions (12a-d) can be solved using a series solution method and additional assumptions related to the physics of the problem. The solution procedure is as follows.

(1) Solve Eq. (11) to find $H(r)$.

(2) Use Eq. (9a) to find $G(r)$ and then obtain $w(z,r) = z G(r)$.

(3) Find pressure drop using Eqs. (9b,c).

In order to solve Eq. (11), a series solution in the form:

$H(r) = \sum_{n=0}^{\infty} a_n r^n$ is used. Substitution of this series solution into Eq. (11) gives:

$$H(r) = a_1 r + a_3 r^3 + a_5 r^5 + a_6 r^6 + ..., \tag{13}$$

$$a_0 = a_2 = a_4 = 0, \quad a_1, a_3 \neq 0$$

$$11 a_1 a_3 + 96 \vartheta a_5 = 0, \quad -6 a_1 a_3 + 525 \vartheta a_6 = 0, \; ... \tag{14a,b}$$

It is seen that the boundary condition: $H(0) = 0$, is satisfied. The function G can then be obtained using Eq. (9a) as:

$$G(r) = -2 a_1 - 4 a_3 r^2 - 6 a_5 r^4 - ... \tag{15}$$

The overall second order form of the axial velocity component should be:

$$w(z,r) = z G = z \left(-2 a_1 - 4 a_3 r^2 + O(r^4) \right) \tag{16}$$

If it is assumed that the parabolic fully developed velocity profile is attained in the microchannel where the injection at the wall ends ($z = L$), and remembering that r is very small in the ranges of micro and nano scales, the high powers in Eq. (16) can be neglected. Therefore, the appropriate form of the axial velocity component will be:

$$w(z,r) = \left(-2 a_1 - 4 a_3 r^2 \right) z \tag{17}$$

The fully developed velocity profile in a cylindrical channel or a pipe is given by [16]:

$$w_f(r) = 2 \overline{w} \left(1 - \frac{r^2}{R^2} \right) \tag{18}$$

where $\overline{w}$ is the average flow velocity. Eq. (16) at $z = L$ should result the fully developed velocity profile given by Eq. (18). The average flow velocity can be obtained from Eq. (7) as:

$$\overline{w}(L) = 2 V \frac{L}{R} \tag{19}$$

Inserting Eq. (19) into Eq. (18) and using Eq. (17) at $z = L$ gives:

$$4 V \frac{L}{R} \left(1 - \frac{r^2}{R^2} \right) = \left(-2 a_1 - 4 a_3 r^2 \right) L$$

The equality of the coefficients in the both side of the above equation gives the coefficients a_1 and a_3 as:

$$a_1 = -2 \frac{V}{R}, \quad a_3 = \frac{V}{R^3}$$

The velocity components: $w(z,r) = z G$, $v(r) = H(r)$ can be obtained as:

$$w(z,r) = G z = \frac{4 V}{R} \left(1 - \frac{r^2}{R^2} \right) z = 2 \overline{w}(z) \left(1 - \frac{r^2}{R^2} \right) \tag{20a}$$

$$v(r) = H(r) = \frac{V}{R} r \left(\frac{r^2}{R^2} - 2 \right) \tag{20b}$$

It is seen that the boundary condition $H(R) = -V$ is also satisfied. Eq. (20a) and its approximate extensions have been previously recommended by Busse [17] for the study of vapor flow in heat pipes. This solution can be called approximate second order solution. However, we can find higher order approximate solutions by taking into account the additional terms in the series solution. For example, a fourth order solution can be obtained by employing boundary conditions $v(r = R) = -V$ or $H(R) = -V$ and $w(z,R) = 0$ or $G(r = R) = 0$ in Eqs. (13) and (15) at $r = R$:

$$a_1 R + a_3 R^3 + a_5 R^5 = -V \tag{21a}$$

$$-2 a_1 - 4 a_3 R^2 - 6 a_5 R^4 = 0 \tag{21b}$$

Eqs. (21a,b) along with Eqs. (14a) provide a set of three nonlinear equations for a_1, a_3, and a_5. In this case, the results show that our problem may have more than one solution. Finally, the velocity components w, v can be obtained as:

$$w(z,r) = -2 (a_1 + 2 a_3 r^2 + 3 a_5 r^4) z \tag{22a}$$

$$v(r) = a_1 r + a_3 r^3 + a_5 r^5 \tag{22b}$$

In this case, the axial velocity profile at $z = L$ can obtained using Eq. (22a).

$$w_f(r) = -2 (a_1 + 2 a_3 r^2 + 3 a_5 r^4) L \tag{23}$$

This solution can be called approximate fourth order solution. The approximate second order and fourth order solutions are compared in the next section.

The pressure drop can be easily obtained by the integration of Eq. (9b) along the microchannel. Substitution of functions G and H into Eq. (9b) and integration from $z = 0$ to $z = L$ gives:

For the second order approximation solution:

$$P(z,r) - P(0,r) = -\left\{ \frac{2\mu V}{R^3} + 2\rho \left(\frac{V}{R}\right)^2 \left[1 - \left(\frac{r}{R}\right)^2 + \frac{1}{2}\left(\frac{r}{R}\right)^4 \right] \right\}$$

(24)

For the fourth order approximation solution:

$$P(z,r) - P(0,r) = -\left\{ 16\mu \left(a_3 + 6a_5 r^2 \right) + \right.$$
$$+ \rho \left[4\left(a_1 + 2a_3 r^2 + 3a_5 r^4 \right)^2 \right.$$
$$\left. \left. + 8 r^2 \left(a_1 + a_3 r^2 + a_5 r^4 \right)\left(a_3 + 3a_5 r^2 \right) \right) \right] \right\}$$

(25)

RESULTS AND DISCUSSION

The saturated water vapor flow in a cylindrical microchannel subjected to uniform water vapor flow injection is investigated here as a test case.

The saturated vapor properties at temperature T_{sat}=373.15 K are: $\mu = 120.3e-7\ Pa.\sec$, $\rho = 0.5974\ Kg/m^3$, $\gamma = 1.3$, $C_p = 2.034\ kJ/kg.K$, $h_{fg} = 2251.2\ kJ/kg$ and $R_g = 0.4615\ kJ/kg.K$. C_p and h_{fg} are heat capacity of the vapor and latent heat of vaporization respectively and R_g is the vapor gas constant. Furthermore, assume a cylindrical microchannel with radius of $R = 100\mu m$, and injection length of $L = 10\ cm$ subjected to two radial Reynolds numbers ($\mathrm{Re}_r = VR/\vartheta$): $\mathrm{Re}_r = 0.6\times10^{-6}$ and $\mathrm{Re}_r = 2.4\times10^{-6}$. It can be easily shown that the maximum Mach number in the microchannel occurs at $z-L$ and is very smaller than 0.3 for both radial Reynolds numbers. The Knudsen number is also slightly smaller than 10^{-3} in both cases. Therefore, the vapor can be assumed to be incompressible and the continuum model or Navier-Stokes equations with no-slip boundary conditions are applicable [18].

Figure 2 shows velocity profiles along the microchannel at two radial Reynolds numbers. It can be seen that as the radial Reynolds number increases the model works well and shows the evolution of the velocity profiles reasonably. Unfortunately, there is no available data in the literature for cylindrical microchannels subjected to uniform wall injection to compare with these results.

Figure 3 shows a comparison between the second and fourth order velocity profiles at $z = 0.05\ m$. Here, there is only one acceptable solution for the fourth order approximation solution at both Reynolds numbers. There is also a small difference between two and fourth order velocity profiles as well. It is important to note that the fourth order approximation

solution does not necessarily represent more accurate results and many additional terms are usually needed to find sufficiently accurate results. This is the subject of the future studies of the author.

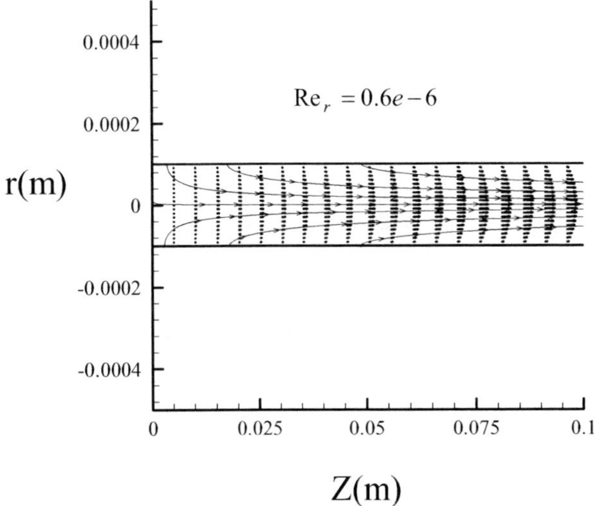

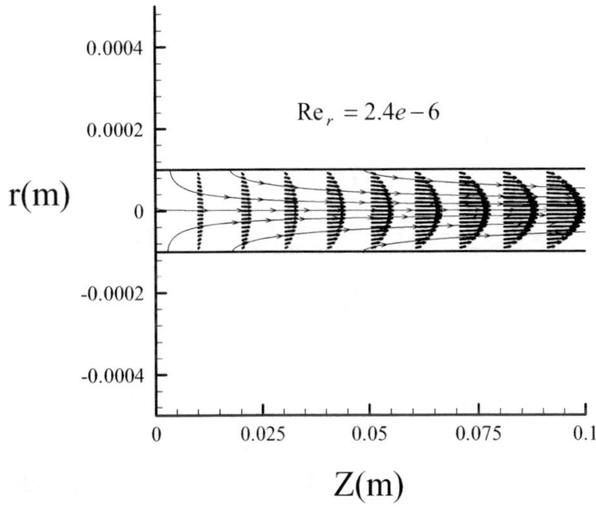

Fig. 2 Velocity profiles in a microchannel subjected to uniform wall injection

Figure 4 shows the corresponding pressure distributions along the microchannel. The pressure at $z = 0$ is assumed to be zero. It can be seen that as the radial Reynolds number increases the pressure drop increases too. The differences between second order and fourth order solutions are also very small.

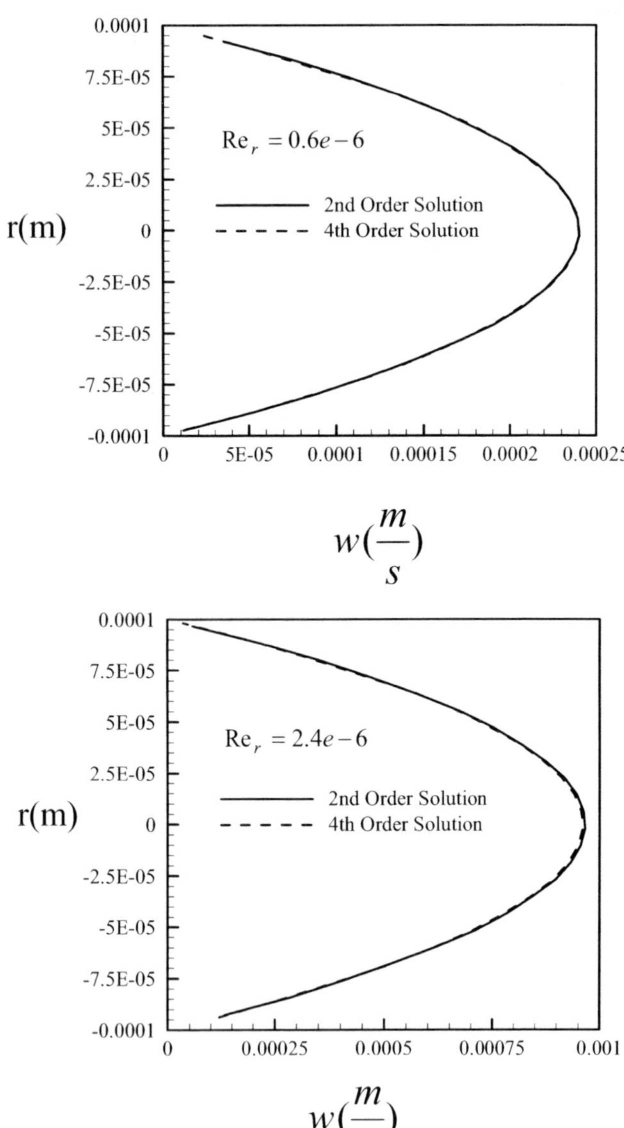

Fig. 3 Comparison between second and fourth order velocity profiles in a microchannel subjected to uniform wall injection

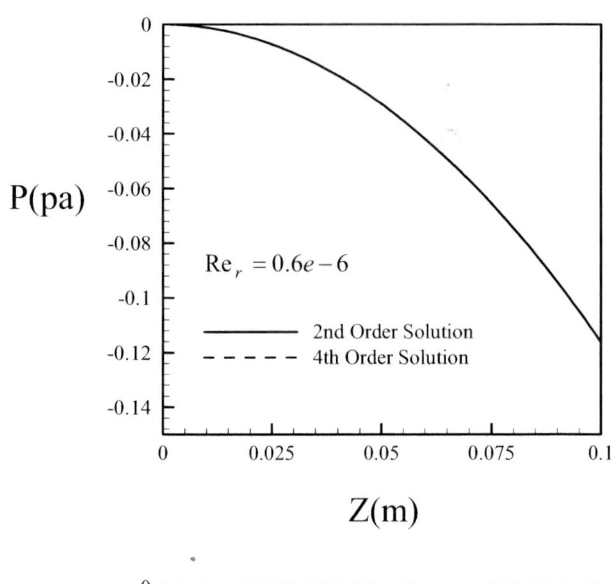

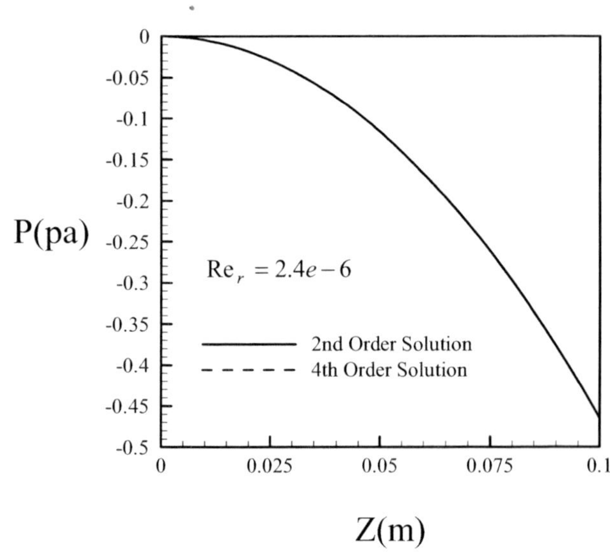

Fig. 4 Pressure drop along a microchannel subjected to uniform wall injection

Figure 5 shows the pressure distributions along a pipe with 1 cm radius and subjected to uniform wall injection, at two radial Reynolds numbers: $\mathrm{Re}_r = 0.6$ and, $\mathrm{Re}_r = 2.4$. It can be seen that as the radial Reynolds number increases the pressure drop increases along the pipe but still remains parabolic. The agreement between the present model and the previous results are also good as shown in Fig. 5. However, as the radial Reynolds number increases, the differences between the second and fourth order solutions increase. It is important to note that although the compared data in Fig. 5 have been given by previous researchers for the vapor flow analysis in heat pipes [17,19,20], but the non-dimensional parameters such as radial

Reynolds number and the assumptions used by the previous researchers are the same as those used in the present analysis.

Figure 6 shows a comparison between the second and fourth order velocity profiles at $z = 0.05\,m$, at two radial Reynolds numbers: $\mathrm{Re}_r = 0.6$ and $\mathrm{Re}_r = 2.4$. It can be seen that the differences between the second and fourth order solutions are more obvious in these cases. Specifically, maximum differences occur at the centerline of the pipe. This is due to the fact that at small scales in the range of micro or nano scales, we can usually neglect the effect of higher order terms in the series solution. This is because the effects of higher order terms on the velocity profiles and pressure drop are negligible as shown in Figs. 3 and 4. However, at larger scales, more terms may be taken into account in order to find sufficiently accurate results.

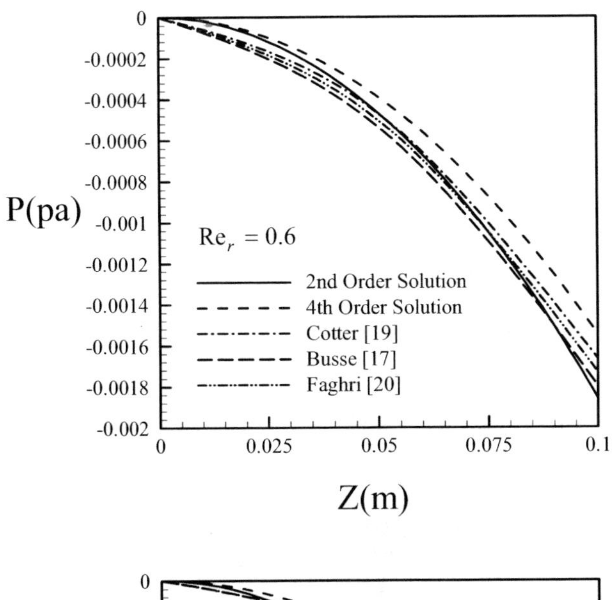

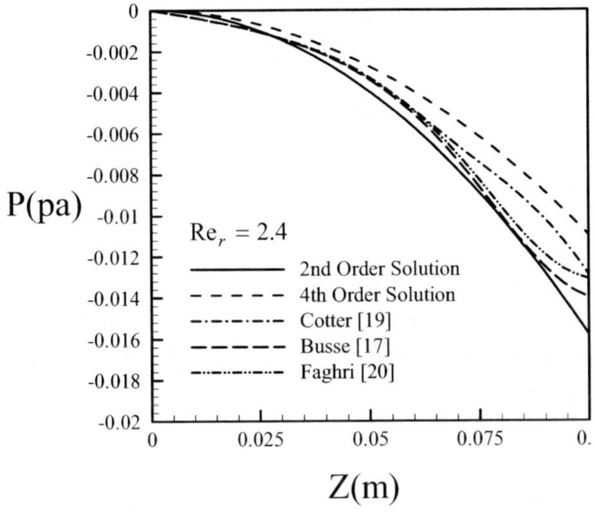

Fig. 5 Pressure drop along a pipe subjected to uniform wall injection

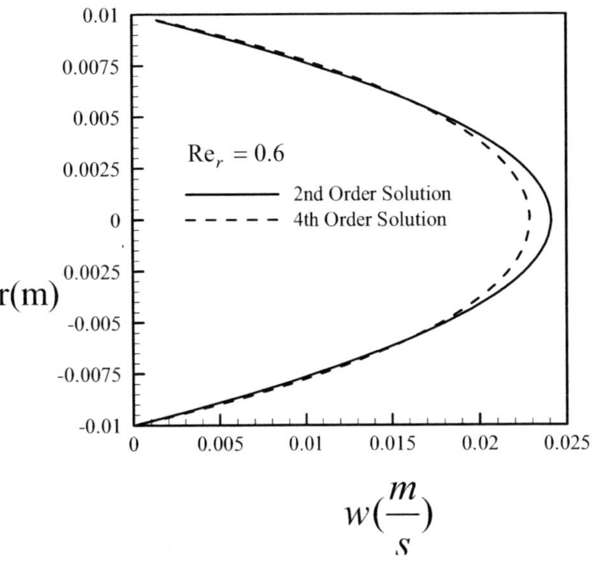

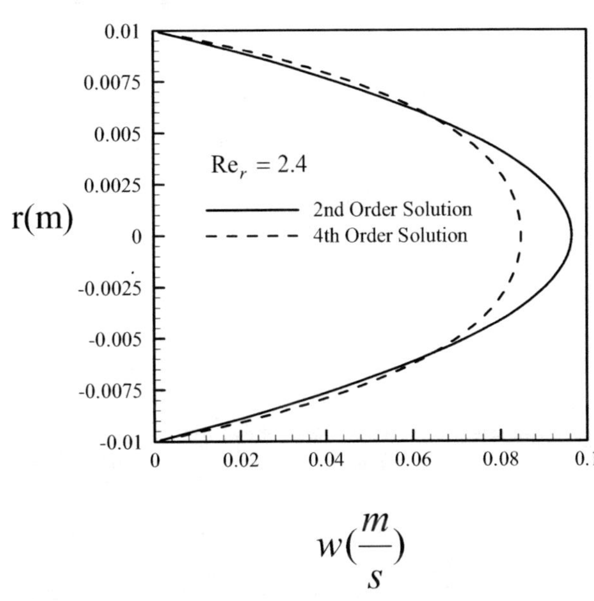

Fig. 6 Comparison between second and fourth order velocity profiles in a pipe subjected to uniform wall injection

CONCLUSION

Analytical analysis of fluid flow in cylindrical microchannels subjected to uniform wall injection at various Reynolds numbers has been presented. The classical Navier-Stokes equations have been used in the present study. Mathematically, using an appropriate change of variable, Navier-Stokes equations have been transformed to a set of nonlinear ordinary differential equations. The governing equations have been solved analytically using series solution method.

It is shown that the laminar and incompressible fluid flow in a cylindrical microchannel can be accurately simulated using the present model. It is also demonstrated that the model works well for the prediction of saturated water vapor flow and pressure distribution in the cylindrical microchannels and presents reasonable results. However, further studies and more complete models are needed to verify the limitations of the present model. The vapor pressure distributions and velocity profiles along the microchannel are predicted for a number of test cases in the range of low to moderate Reynolds numbers ($Re_r \leq 2.4$). The results of the presented analytical model have

145

been compared with the available numerical data in the literature and have shown good agreement in all cases. The present analysis can be easily extended to a more complete analysis of micro or nano heat pipes. This is the subject of the author's current research.

ACKNOWLEDGMENTS

I would like to thank Professor Kambiz Vafai at the University of California, Riverside, who encouraged me to study about the interesting concept of fluid flow in microchannels.

REFERENCES

[1] Stone, H. A. and Kim, S., 2001, "Microfluidics: Basic Issues, Applications, and Challenges," AIchE Journal, **47**, pp. 1250-1254.

[2] Jensen, K. F., 1999, "Micromechanical Systems: Status, Challenges, and Opportunities," AIchE Journal, **45**, p. 2051.

[3] Langer, R., 2000, "Biomaterials: Status, Challenges, and Perspectives," AIchE Journal, **46**, p. 1286.

[4] Happel, J. and Brenner, H., 1983, *Low Reynolds Number Hydrodynamics*, Martinus Nijhoff.

[5] Kim, S. and Karrila, S. J., 1991, *Microhydrodynamics*, Butterworth-Heinemann.

[6] Gallardo, B. S., Gupta, V. K., Eagerton, F. D., Jong, L. I., Craig, V. S., Shah, R. R., and Abbott, N. L., 1999, Electrochemical Principles for Active Control of Liquids on Submillimeter Scales", Science, **283**, p.57.

[7] Chang, H. C., 2001, *Bubble/Drop Transport in microchannels*, CRC MEMS Handbook, M. Gad-el-Hak.

[8] Sharp, K. V., Adrian R. J., Santiago, J. G., and Molho, J. I., 2001, *Liquid Flow in Microchannels*, CRC MEMS Handbook, M. Gad-el-Hak.

[9] Arkilic, E. B., Schmidt, M. A., and Breuer K. S., 1997, "Gaseous Slip Flow in Long Microchannels," J. of Microelectromech. Sys., **6**, p.167.

[10] Pit, R., Hervet, H., and Leger, L., 2000, "Direct Experimental Evidence of Slip in Hexadecane: Solid Interfaces," Phys. Rev. Lett., **85**, p.980.

[11] Rahman, A., 1964, "Correlations in the Motion of Atoms in Liquid Argon," Phys. Rev. A, 136, pp. 405-411.

[12] Allen, M. P., and Tildesley, D. J., 1987, *Computer Simulation of Liquids*, Clarendon Press, Oxford.

[13] Haile, J. M., 1992, *Molecular Dynamics Simulation: Elementary Methods*, Wiley, New York.

[14] Rapaport, D. C., 1995, *The Art of Molecular Dynamics Simulation*, Cambridge University Press, Cambridge, UK.

[15] Allen, M. P., 2004, *Introduction to Molecular Dynamics Simulation*, NIC Series, Vol. 23, ISBN 3-00-012641-4, pp. 1-28.

[16] White, F. M., 1991, *Viscous Fluid Flow*, McGraw-Hill, New York.

[17] Busse, C. A., 1967, "Pressure drop in the vapor phase of long heat pipes," *Proc. 1967 IEEE Thermionic Conversion Specialist Conf.*, Palo Alto, California, p. 391.

[18] Gad-el-Hak, M., 2001, "Flow Physics in MEMS," Mec. Ind., **2**, pp. 313-341.

[19] Cotter, T. P., 1965, "Theory of Heat Pipes," Los Alamos Scientific Laboratory, Report No. LA-3246-MS.

[20] Faghri, A., and Chen M. M., 1989, "Numerical analysis of the effect of conjugate heat transfer, vapor compressibility and viscous dissipation in heat pipes," Numerical Heat Transfer, **16**, pp. 398-405.